RESOURCES, ENVIRONMENT, AND ECONOMICS

Resources, Environment, and Economics

Applications of the Materials/Energy Balance Principle

ROBERT U. AYRES

A WILEY-INTERSCIENCE PUBLICATION

JOHN WILEY & SONS, New York · Chichester · Brisbane · Toronto

Library of Congress Cataloging in Publication Data:

Ayres, Robert U.
  Resources, environment, and economics.

  "A Wiley-Interscience publication."
  Includes bibliographical references and index.
  1. Econometrics. 2. Economics--Mathematical models.
I. Title.
HB141.A97   330'.01'82   77-20049
ISBN 0-471-02627-1

Printed in the United States of America

10 9 8 7 6 5 4 3 2 1

Preface

The purpose of this book is to acquaint the wider community of academic and institutional resource economists and environmental systems analysts—that is to say, those who are not directly involved in large-scale simulation modeling (and even, perhaps, some who are)—with some recent developments in this field that for the most part have not yet been published in the journals.

The book is intended as a monograph, not a text for teaching purposes, although I am bold enough to hope it will be seen by at least some graduate students. It attempts to explain some of the new models and relate them to the traditional concerns of economists. However, no attempt at an absolutely rigorous mathematical presentation is made.

In the first place, I was trained in mathematical physics, not mathematical economics, and am not particularly comfortable with the language and approach that are second nature to the most sophisticated of that breed. Hence it seems wise to make a virtue of necessity and not to attempt a second-rate imitation of the style of John Von Neumann, T. C. Koopmans, or Kenneth Arrow.

In the second place, large-scale models are designed and used for very different and less rigorous purposes than the small models so common in the economics literature. The key differences will be discussed and stressed at many points in the book. Briefly, however, small aggregative models in economics are developed for purposes of generating *theorems*, which are precise statements about the relations between various factors under carefully specified conditions. Often the specified conditions are far from realistic, and the practical value of each exercise is slight until a great many simplified models, based on different assumptions, have been examined and the results compared. Even then, the truly generalizable statements are rare—and always subject to modification as a result of analysis of the next such small model. Regrettably, academic economists not infrequently generalize too freely from the results of ultra-simplified models. A theorem based on restricted assumptions is gradually transformed over the years into a "law" or a "principle" (e.g., Hotelling's law). Careless habits of thought can, and sometimes do, mislead both professional economists and policy-makers who accept their advice. One or two examples are cited in the text from the literature of resource economics and environmental economics.

Obviously large-scale models cannot and are not expected to serve the same intellectual purposes as simplified ones. Realism is their *raison d'être*. Of course, to achieve realism in one area often involves some sacrifice of realism in others. Thus realistic models often do not have properties that are theoretically desirable. Moreover, realistic models are necessarily data limited. Abstract models can freely introduce nonmeasurable quantities (such as consumer utility or public welfare) but realistic ones cannot. Finally, abstract models need not be "solved" in practice. Only the general properties of the solutions are of interest and these can generally be deduced much more easily than actual numerical results.

Realistic model-builders necessarily spend much of their time on acquiring and processing data, and on developing (and correcting) computerized algorithms to solve the equations. Abstract theorists can avoid these diversions. To them the actual numbers matter little in any case. Why, then, should a theorist concern himself or herself with the properties of very large and complex models? There is only one reason. Economics is, after all, a *social* science, meaning that it is ultimately concerned with aspects of human behavior. Even the most abstract economist must come to grips with reality occasionally to see whether he is asking (or answering) the right questions. The large-scale model-builder is constantly face-to-face with this imperative. It is, so to speak, his bread and butter. And, notwithstanding the many distractions inherent in model-building, any insights this book may contain as to what new phenomena are important and relevant, may yet be of some modest value to the theoreticians still residing in the ivy-covered ivory towers.

Another purpose of writing this book is to address some methodological issues to the econometrics subset of the economics community. The term econometrics connotes the application of quantitative *measurement* to economics and, by extension, the formulation of quantitative relationships. In short, econometrics is, properly speaking, the empirical branch of economics. In principle, a variety of methodologies should be fruitful, depending on the nature of the problem. Unfortunately, only one rather narrow approach has been applied intensively: the use of multiple correlation analysis of statistical data.[†] To quote one of the leading practitioners, Lawrence R. Klein,[‡]

In large part, econometrics became an application of multiple correlation theory. There is more to econometrics than the estimation of relationships, but that single aspect of the subject has swamped everything else.

[†]Like many nonstatisticians, I suspect, I have a sloppy tendency to use the terms correlation and regression more or less interchangeably. Actually there is a difference: Correlation does not imply causality, whereas regression does. As a tool for forecasting time series, regression is regarded as a bad estimation procedure.

[‡]R. L. Klein, "Whither Econometrics?", in Rendigs Fels and John J. Siegfried (Eds.), *Recent Advances in Economics*, Richard Irwin, Homewood, IL., 1974.

I have no objections at all to the judicious use of statistical methods where they are appropriate, nor do I doubt that multiple correlation (and regression) methods are the most appropriate tools in a great many of the situations where they have been applied. But they are *not* appropriate in all cases, and examples are fairly easy to cite where the standard econometric approach as applied by a statistician with no substantive knowledge of the underlying social and physical realities would almost certainly end up—after immense effort—with estimating equations that are transparently incapable of making a credible forecast (or, perhaps, even with equations that violate physical principles).

The point that needs to be emphasized is that there are sometimes other types of information available—frequently from the physical or biological sciences—that are grounded in basic principles governing the relationships of matter and energy. These principles may not comprehend ultimate Scientific Truth, but they should and must supersede purely statistical inferences.

If economics were only concerned with the exchange of abstract services, these remarks would be less relevent. In point of fact, however, services are often generated by materials goods, which are manufactured from physical commodities. By the same token, many disservices arise from these same physical materials as their "utility" is exhausted and they become wastes. Moreover, economics is intimately concerned with optimizing production, which means that "production possibilities" must be delineated. Once again, the properties of real materials and energy, and the attributes of real machines, processes, and products are necessarily involved.

To make the essential point clear, one might suppose that an econometrician using only the tools of multiple correlation analysis were asked to forecast the future fossil fuel requirements (in Btu) for generating electricity, based on statistics for generating plants built between (say) 1870 and 1960. Would he predict correctly for the year 2000? The answer is, almost self evidently, *no*. However much the statistical estimating procedure is disguised by fancy wrappings, it yields (in this case) nothing more than a "best fit" extrapolation of some declining function of time. It is to be hoped that the econometrician would know enough to discard functions that could become negative—implying that sometime in the future, power plants will "breed" fuel instead of consuming it. But the knowledge in question is derived from physical laws: It cannot be inferred from the historical time-series data. And, unless the econometrician had also been trained in physics, chemistry, or engineering, there is no reason to expect that he would also impose the much stronger, but absolutely essential, constraint that the fuel input must always be *greater* than the electricity output![†]

As I commented earlier, I am not arguing that econometricians should discard statistical multiple correlation (or regression) analysis techniques. Not at all. I am, however, arguing for a more explicit recognition of the role of nonstatistical information, whether theoretical or empirical and whether derived from other disciplines or from other branches or economics.

This book is intended, in part, as a first step in showing where, and how, this interdisciplinary fusion can be accomplished in the construction of large-scale "realistic" economic models for policy analysis.

ROBERT U. AYRES

Washington, D. C.
January 1978

[†]Both are measured in the same units, of course. The constraint is, in other words, that conversion efficiency can never, under any circumstances, exceed 100%.

Acknowledgments

For my introduction to resource and environmental economics, and for my early tutoring on the subject I would like to thank Allen V. Kneese, formerly Director of the Equality of the Environment Program, Resources for the Future, Inc., Washington, D.C., and now in the Department of Economics, University of New Mexico, Albuquerque, New Mexico. For continuing education and guidance into the subtleties and intricacies of input–output models, I would like to thank Stedman Noble. For substantial contributions and key ideas as well as much of the implementation of the materials–process–product model, particular credit is due to Martin Stern, James Cummings-Saxton, and Robert Roig. These individuals, all former colleagues at International Research and Technology Corporation, are listed as honorary co-authors, respectively, of Chapters 4 and 5.

Other individuals, past and present, who contributed substantially to various parts of the work which is summarily reported here, include Ivars Gutmanis and Adele Shapanka who contributed to the early work on extending the input–output model to incorporate waste flows. More recent work along these lines, particularly to disaggregate the input–output model to the level of materials and commodities, has been carried out primarily by Marc Narkus-Kramer, Richard Meyer, and Andrea Watson. A large input–output table for the chemical industry was developed under the direction of Dr. Cummings-Saxton by Albert Loridan and Nickolaus Leggett. The analysis of direct and indirect energy flows using the input–output model has been carried out by Tom Betlach, Craig Decker, Marc Narkus-Kramer, Richard Meyer, and Ralph Doggett. Appendix 6A of Chapter 6 was compiled by Eric Weinstein. Computer programming has been primarily done by Hillel Sukenik, Leslie Ayres, and Beverly Moore.

Credit is also due to a number of long-suffering clients and sponsors who have supported research projects of IR & T from 1970 through 1975. In rough chronological order, these include Resources for the Future, Inc.; the National Science Foundation; the Bureau of Labor Statistics; the Department of Transportation; the Environmental Protection Agency; the National Commission on Water Quality; the Energy Research and Development Administration; and the Office of Energy Conservation, Energy, Mines and Resources, Canada. Individuals in these organizations are almost too numerous to list, but I must pay special tribute to Ronald Ridker at Resources for the Future, Inc.; Joseph Coates at National Science Foundation; Peter House at the Environmental Protection Agency; James Motley and Philip Patterson at the Energy Research and Development Administration; and David Brooks at EMR, Canada.

Finally, thanks are due to Simon Goldberg, Nancy Ruggles and Keith Walton of the United Nations Statistical Office in New York; Mark Jansen, of the Statistical Division, Economic Commission for Europe, Geneva; and Anthony Friend and Bert McInnes of Statistics Canada, for their help in developing the conceptual basis for a System of Environmental Statistics, which is discussed in the last chapter of this book.

Professor Ralph d'Arge, Department of Economics, University of Wyoming, deserves my special gratitude for providing me with a "hideout" both to get this book started away from office distractions, and again later on, to finish it. He also provided invaluable criticism and necessary additional tutoring in the methods of abstract economic theory, plus a willing ear and much encouragement. Others who have reviewed part or all of the manuscript and made very helpful suggestions or criticisms include Robert Dorfman of Harvard University, Fred Peterson of the University of Maryland and the Council of Economic Advisors, and Charles A. S. Hall of the Marine Biological Laboratory, Woods Hole, Massachusetts. Very much thanks are due to my secretary, Frances Calafato, and to Barbara Stanin, Sharon Walker, Caroline Norberg, Martha Yopes, and Lucile Stafford at IR & T. I would also like to thank my wife, Leslie Ayres, for more kinds of help and support than can be described in words.

R.U.A.

Contents

RESOURCES, ENVIRONMENT, AND ECONOMICS

CHAPTER ONE *Economics and the Environment*

Pure science can be said to be stimulated by human curiosity about the basic way in which things work—the need to understand for the sake of understanding. There is at best a narrow line, which sometimes vanishes altogether, between the generalization required to understand and the simplification that facilitates manipulation of abstractions. Applied science, on the other hand, is attributable largely to the need to solve problems. In economics, which is after all concerned with some of the most fundamental aspects of human relationships and behavior, the distinction between "pure" and "applied" is often particularly blurry, even to the practitioners. Indeed, most work on economics has had relatively immediate applications, at least until very recent times. Lately, however, theory and application in some branches of economics have begun to diverge to some extent, with unfortunate results. This tendency is especially noticeable in the realms of environmental and resource economics. The work reported in this book is motivated by a recognition of the need for better tools to manage our tangled affairs, which can only come from a reconciliation of theory with the imperatives of the real world.

The analogy between the 1970s and the 1930s is interesting in terms of the development of economic models. The great stock market "crash" of 1929 followed by the Great Depression of 1930–1939 brought into focus a number of issues that had been simmering for a long time in the economics profession. Perhaps the most important one was the origin of the "business cycle" in terms of a detailed interplay among production, consumption, investment, savings, and taxation.[1-3] The great contribution of John Maynard Keynes[4, 5] was to develop these concepts into a *policy* model. In particular, Keynes postulated a relationship of algebraic form between personal income and consumption expenditures with the balance, of course, going to savings and taxes. This basic theory led to the first mathematical models of the business cycle in the 1930s.[6, 7]

The depression, according to Keynes, resulted from the fact that savings were not immediately translated into investment goods, because increased capacity was not justified by the low level of consumer demand. As a result, substantial resources, both of labor and capital, remained unutilized. Keynes developed an important policy recommendation out of this analysis, mainly that government should act as a counterbalance and create additional demand—if necessary via deficit spending—doing more public works projects and the like. On the other hand, during periods of high consumer demand and high employment, the government should not compete with consumer demand, but rather adjust its expenditures to a level below revenues, thus creating a budgetary surplus to balance the antirecession deficit. This policy recommendation has been extremely influential in the management of national economies in recent decades, even though the prescription has never been followed fully.

1.1 THE ECONOMY AS A PRODUCTION–CONSUMPTION MACHINE

Classical economic theory considers the economy to be a kind of perfect ageless machine. In this view, there is a cyclic relationship between *production* or supply on the one hand and *consumption* or demand on the other. The "working fluid" of this machine is money, and money (like materials or energy) is subject to an accounting identity that is equivalent to a kind of conservation principle: For any non-governmental economic entity, such as a household or firm, money inflows (income) must always be exactly balanced by outflows (expenditures) plus any changes in stock (savings or reserves). Governments, of course, can create new money by various mechanisms that complicate the picture, but this can be neglected for the moment.

A key insight worth emphasis because of its great importance later in this book, is the closed "cyclic" nature of the economic system. Thus Gross National Product (GNP) is defined as the sum of all outputs by all producers (i.e., sellers) in the economy. But the sum of all outputs is, by virtue of the accounting identity, equal to the sum of all expenditures by all consumers (i.e., buyers) in the economy. Thus Gross National Product is, necessarily, equal to Gross National Expenditure.

The duality between production and consumption also translates into a duality between "costs" on the one hand and "prices" on the other. The cost of producing a given bill of goods (including services) consists of all payments for purchased materials, capital equipment, interest on loans, and dividends on invested equity capital, plus taxes and wages. Outlays for purchased materials and capital equipment are, in turn, attributable to the same categorical elements, in various proportions. Ultimately, all payments for products and services can, therefore, be attributed to wages, rents, royalties, or other types of income to individuals.

Personal income, in turn, is allocated to savings, taxes, and private final consumption expenditures. The other categories of final consumption are government (public) consumption and investment (both private and public). Investment differs from consumption in that it is intended to increase the

capacity for production of other goods and services.[†] Private savings may or may not be reinvested in producer goods by financial institutions. Some lending goes to consumer credit, and some goes to the government. Consumption and investment expenditures for goods and services, in turn, provide the revenues (income) for producers. Revenue is, of course, the product of quantity times price, so prices are "costs" for consumers just as wages are "costs" for producers. This is the reason that raising money wages without increasing physical production ultimately increases price levels. Raising interest rates, incidentally, increases both producer and consumer costs and consequently—other factors remaining the same—tends to reduce both production and consumption, while increasing savings.

The government "loop" involves taxes collected from both consumers and producers, which constitute income for the government. Government expenditures constitute the reverse of this flow. First, there are "transfer payments" such as welfare, social security, medicaid, agricultural subsidies, and so forth, constituting personal income for certain individuals. Second, there are salaries and wages to government employees. Third, government expenditures contribute to demand for some kinds of goods and services produced by the private sector, notably military equipment, public buildings, and roads. All of this is complicated by the fact that government, unlike producers and consumers, need not live with any kind of budget constraints if it is not deterred by the risk of inflation. Thus, whereas both consumers and producers are limited in their buying and investing capacity by current income and savings (which determine borrowing power), government has an effectively unlimited capacity to borrow by increasing the money supply and inflating the currency. Government also controls interest rates and tax rates, of course. To complicate matters, there is no central direction for Western mixed economies, nor do economists (or politicians) possess accurate knowledge of production functions, consumption functions, supply constraints, technological options, price elasticities, or "social" choice tradeoffs. Decisions are typically made with slight information and vast uncertainty.

[†]In practice, a further reason for distinguishing between consumption and investment goods is that investment goods are also longer lasting than consumption goods on the average (except private housing, which could also very well be regarded as a form of investment).

1.2 PRODUCTION–CONSUMPTION MODELS

These complexities make it very difficult for even the most experienced economists to foresee the consequences of any given policy, still less to design a policy that will yield optimal results, or even to define an optimal outcome. The perceived need to "manage" the economy better has thus inevitably led to a demand for more and more detailed models of the flows of money and goods and services among and between the various key sectors of the economy, as a function of unemployment, unused industrial capacity, interest rates, and so on. To be of use, these models had to distinguish between various constituent elements of final demand; for example, consumables, consumer durables, producer durables, and so on. It was necessary to distinguish different industrial sectors, such as mining, manufacturing, retail and wholesale, utilities, and transportation. Imports and exports of goods and services had to be taken into account as well as transnational flows of money for capital investment or intergovernmental loans. Interest rates had to be taken into account, changing price levels had to be considered, along with the effect of taxes. Thus the 1930s saw the beginning of the first serious efforts to build aggregated mathematical models of a national economy, by Jan Tinbergen[6, 7] and Lawrence Klein[8] among others.

Input–output models, in the modern sense, also began to evolve in the 1930s, although clearly anticipated by F. Quesnay as far back as 1766.[9] The initial impetus seems to have come from the USSR,[10] which was trying to construct an industrial society from the "top down" and needed analytic tools to assist the central planners. Wassily Leontief's first description of a model exhibiting general interdependence between major factors of production at the sectoral level was published in 1937.[11, 12]

As the need for models was recognized, a further need became immediately clear, namely the need for much more detailed and more reliable information on these monetary flows and accounts. Thus was born the concept of National Accounts. This was another area where Simon Kuznets,[13, 14] Richard Stone,[15] and others made great contributions. Since National Account data is intrinsically international because of foreign trade and capital flow components, it quickly became clear that an international system of National Accounts was required. This problem was taken up in the early years of the United Nations and was eventually developed into an international System of National Accounts (SNA),[16] developed primarily by Professor R. A. Stone at Cambridge University. Over the past 20 years a relatively reliable and comprehensive system of national accounts data has been developed not only for the United States but also for most of the major trading countries of the world, and with the help of this data base, a whole new generation of econometric and input–output models has evolved.[†] This is not to suggest, of course, that the present state of either the models or the statistical underpinnings is considered satisfactory, either by the economics profession itself, or its "clients."

The purpose of the preceding brief discussion of traditional economic concerns was to emphasize the essential relationship between the recognition of theoretical inadequacies in economics, and governmental inability to deal with the economic problems of the time—mainly the Great Depression of the 1930s, not to mention the catastrophic inflations of the previous decade—and the subsequent development of economic models. A comparable link can be traced between the increasing demand for models and the development of statistical data bases both nationally and internationally to facilitate the implementation of those models. Now some 40 years later, we are approaching the ability to use these production–consumption models constructively in managing the economy. Indeed, in moments of euphoria, economists are even beginning to speak of

---

†Some of the major econometric growth models are the Wharton model,[17] the Brookings model,[18] the Federal Reserve–MIT model,[19] the Michigan model,[20] the Harvard–BLS model,[21] and the Chase Econometrics model.[22] Similar models have been developed in recent years for many other countries, and an international meta-model known as LINK has evolved, which ties together national econometric models for some 20 major trading countries.[23] This is the beginning of a truly international forecasting and simulation system, capable of handling the whole range of economic and monetary problems of major concern to governments today. The most widely used input–output forecasting models in the United States are those of the University of Maryland[24] and the Bureau of Labor Statistics.[25] Also, in recent years, input–output models have begun to transcend national borders, and a consistent set of international input–output tables has been developed under the direction of the Economic Commission for Europe (ECE), and an international regional-level input–output forecasting model has been built by Professor Leontief and several colleagues, and applied to some key questions relating to the impact of environmental and development problems on economic relations between major trading nations.[26]

"fine tuning" as a practical possibility. In all fairness, the recent wave of criticism of econometric models for their failure adequately to account for, or predict, the inflationary and recessionary consequences of recent sharp increases in petroleum prices and other events of the last few years, should not be taken as a total indictment of the capabilities of such models. To a very large extent, our notion of what is a satisfactory level of accuracy to be expected has evolved in step with the models themselves so that we are almost as dissatisfied with the state of our models today as we were in the 1930s, although the comparison between the models of the two periods is really like comparing a Ford Trimotor with a supersonic Concorde jetliner.

1.3 THE NEED FOR GENERALIZATION OF THE BASIC PARADIGM

The 1960s and 1970s have introduced a new set of characteristic problems. Not that the problems of the 1930s have been "solved." There is yet much to be done to adapt the new production–consumption modeling capabilities to the practical and political difficulties of managing a modern industrial economy. But economists do not seem to be aware of major theoretical gaps in that area of economics and the deficiencies of the models seem to be pretty much in the area of data acquisition and "fine structure." The emerging problems of the 1970s for which our theory is much less complete, and the data base correspondingly so, are the problems that are associated with two hard and uncomfortable facts. First, the economy does not operate on abstractions. Second, the old, convenient assumption that the economy operates at (or, at least, tolerably close to) a state of Pareto optimality in which marginal private costs are always equal to marginal social costs, is invalid. In reality, the deviations between private and social costs are frequently significant.

Production and consumption are intimately involved with *real* physical materials and energy. These real physical materials must be derived originally from the natural environment where they are distributed in a complex pattern. Extraction is not a matter of simple removal, but of highly sophisticated technologies evolving rapidly in time. Materials usage involves successive stages of processing and transportation that inevitably result in social costs (externalities) ranging from noise to the discharge of toxic or obnoxious waste materials. Similarly, so-called "final" consumption—that convenient abstraction from classical economics—is not final at all in terms of disposition of real materials and energy. On the contrary, consumption of material goods means in practical terms that the goods have lost their utility value and become wastes which still have to be disposed of, and which are still capable of causing very serious harm to the natural environment, depending upon the location and method of disposal.

In the 1930s (and earlier) the problems associated with sources of materials and disposal were perceived to be of very small magnitude in comparison with the problems of organizing the economy and fully utilizing its resources.[†] Today the problem-set is shifting. Natural resources, in particular, are becoming painfully scarce within the territorial boundaries of the most heavily industrialized countries of the world. Many of the highest quality natural resources of the Earth have already been used up, not to say wasted. And those countries that still possess stocks of high-quality resources are demanding sharply higher prices and vastly greater control over their exploitation. The surface of the Earth still has abundance to offer, to be sure, but we must dig deeper, scrape the bottom of the ocean, or look at more remote areas of mountains, deserts, and jungles. Extracting these resources is creating newer and more serious problems also. Delicate ecosystems such as the Arctic tundra or the tropical oceans are beginning to show adverse affects from this intensification of digging, drilling, and quarrying. Mine wastes, oil leaks and products of combustion of a vast variety of types and kinds of pollution are fouling our environment and making survival impossible for many harmless species of plants and animals that formerly shared the Earth with us. Some have even questioned whether human life itself can survive for long amidst this carnage. Immediate fears of resource exhaustion or "pollution death" may or may not be overblown, but that is not the issue here.

Part of the burden of this book is to make the point that the "new" problems facing economists in the 1970s and 1980s are intimately associated with the physical quantities (stocks and flows) *and qualities* of materials and forms of energy. Energy is

---

[†] Of course, this is still the major problem of the "less developed" countries, which accounts for the growing divergence of interests between North and South.

required to process materials to manufacture goods or to produce services (such as illumination or environmental conditioning) from goods. Materials are incorporated in goods. But characteristic *disservices* are also associated with all these processes, ranging from destruction of the land surface at the point of extraction to generation of energy or material waste residuals at the point of processing or final consumption. The existence of these disservices is inherent in the use of materials and energy.

The conventional paradigm of economics addresses the economy as a set of relationships between production, investment, and consumption expressed in monetary terms, and defines its concerns as determining conditions for maximizing consumer utility and social welfare by optimizing these relationships. By assuming a Pareto-optimal (or near optimal) state, it is presumed that a partial equilibrium type of analysis is adequate.

My thesis, here, is that a more general paradigm is needed, in which the economy is viewed as a set of transformations of physical materials from the raw state through successive stages of extracting and processing to goods and services, and finally to return flows consisting of wastes (which may or may not be recyclable). The problem of optimization is correspondingly broadened. This broader theory must address the problem of production of externalities as well as economic services, and the allocation of such externalities in a general equilibrium context. It must deal with the problem of defining and maximizing social welfare subject to resource supply constraints, laws of thermodynamics, and the existence of pervasive externalities resulting from waste residuals; and it must provide theoretical tools to facilitate our understanding of the appropriate mechanisms for managing the economy.

Economists are used to thinking of their subject as a *social* science; so it is, insofar as economics concerns itself with human behavior, as it is expressed in the relationships between buyers and sellers of goods and/or services. Actually, since Irving Fisher[27] and Frank Knight,[28] it has been recognized that services are more fundamental since material goods are "consumed" only for the services they provide. Thus it is quite natural for modern economists like Kelvin Lancaster[29] to abstract more and more away from material goods, considered as such, to their basic attributes.

But there is no mechanism by which the attributes of materials can be sold, purchased, or consumed independently of each other or of the materials as such. The color, flavor, and bouquet of wine cannot be disassociated in reality from its liquidity, weight, volume, alcoholic content, and—yes—the dregs in the barrel or bottle. While the mix of attributes may change over time or from one sample to another, there are no materials with only one attribute and very few attributes that can be added to other materials via a highly concentrated "essence." The obvious (partial) exceptions are certain flavoring agents, such as salt or sugar, and certain pigments and dyes. It is meaningless to speak of such qualities as an "essence" of hardness, tensile strength, incompressibility, fluidity, thermal conductivity, or plasticity. Nevertheless, physical materials are used for economic purposes because we value their attributes—at least in certain combinations and circumstances. The inherent linkages between attributes and materials is very important for economics, and central to this book.

It is not only manufactured products (goods) whose specific attributes economists must now consider, but natural ones. Nowadays it is quite clear—though perhaps once it was not—that there is no fundamental distinction in kind between the "satisfaction" or "utility" obtained from goods and services purchased in a marketplace and the satisfaction or utility obtained from nature through nonmarket channels—except the monetary valuation of the marketplace itself. But the absence of a monetary transaction does not mean an absence of value. Obviously a fish is no less nourishing or delicious if it is consumed directly by the fisherman than if it is sold (or bartered) by the fisherman to his neighbor. The same point can be made for any object that is found complete in nature and used as such (not sold) by the finder; or any product that is manufactured by the end-user from "found" raw materials using his own labor or that of unpaid family members. In fact, very important private goods and services are regularly exchanged without any market valuation. In some cases the lack of a market exchange is merely accidental, and an (approximate) equivalent market price is relatively easy to estimate—as with the value of unpaid labor.

Still, classical economics is so market- and money-oriented that it has largely neglected unpurchased and unpriced goods and services even where these commodities are exchangeable in principle and a price could be estimated approximately. Our conventional monetary measures of income and wealth, for instance, are largely irrelevant to rural undeveloped countries (such as India) because a large fraction of the population consume homegrown food, use homemade tools, live in homemade houses, and, of course, neglect to pay themselves in cash for their own labor or that of children and wives. Only when they sell a bit of surplus food or an animal, or buy some cloth or a bullock cart, do they enter the realm of the marketplace—literally.

In the towns and cities of industrially developed countries, people normally sell some (though not all) of their labor for money, and they buy their food, clothing, shelter and furnishings, and many other services, from specialists. Accordingly, economists have long pretended that the goods and services that were and are *not* bought and sold in the market must be relatively unimportant. Not until comparatively recently did economists become aware of the extent to which this picture is oversimplified because it neglects the fact that some kinds of goods—and especially services—*cannot* be exchanged in a market and hence, cannot be given a market valuation (or price).[30-32]

The distinction between ordinary economic goods or services and unpriced or "public" ones may be attributable either to institutional arrangements—notably the laws governing ownership and exchange—or to the inherent physical nature of the good. "Indivisibility" is the usual criterion. For example, the ocean or the atmosphere are said to be indivisible because one cannot reserve, isolate, or "contain" a meaningful chunk of either. Therefore, they cannot be separately owned or exchanged.

For similar reasons, nobody can really "own" a distant view or a raincloud or the sun or the moon or an asteroid or a star. It is possible to buy and sell quite exotic caged animals, but not (by definition) "wild" ones. Nobody owns the birds or the bees or the fish in the sea, until they are caught.

Another characteristic of some goods that interferes with ownership or exchange is that the amount that exists is effectively indefinite. This applies, for example, to the case of the distant view already mentioned. Any number can enjoy it, yet each viewers' enjoyment is unimpaired by the fact that it is shared with others. Technical information is another example. In this case, the difficulty of applying conventional economic reasoning is that, while those who have information can *sell* it (under some circumstances), the seller does not lose the use of

what he sells. The total supply is infinitely elastic; indeed, it increases with each such transaction. The market value of information, if any, is related to its degree of exclusivity. As soon as technical information is widely published it becomes a free common property resource, thence a public good.

A major category of "priceless" services comes from nature. Perhaps the first of these is human life itself, and the associated satisfaction of conscious awareness. Life is said to be a gift, and the description is apt. Certainly life is not allocated on the basis of ability or willingness to pay for it. Life is "free" to each human—as well as to all other organisms—but its quantity is limited by nature. Up to the present time, at least, no amount of wealth could purchase an extension of life beyond its biblical span of four score and ten years or so. With minor and unimportant exceptions, the individual human body and its components cannot be bought or sold; it is essentially indivisible.

Nature also uniquely provides virtually all the requisites for sustaining life on this planet. While Man may be the highest—or at least, the latest—product of biological evolution, human life is supported by a pyramid of other forms down to the lowliest single-cell algae or bacteria. Man's technology and his industry can replace only a few of the elements of the natural ecosystem—such as synthetic chemical fertilizers and pesticides—and only to a limited extent. It is by no means clear that even this degree of human interference in critical natural cycles can be tolerated indefinitely without an ecocatastrophe or collapse of some sort. Certainly a number of respected "systems" ecologists—one might mention Commoner,[33] among others—have expressed doubts on this point.

Because this book is concerned mainly with environmental and resource economics, it is worthwhile to note in some detail just how comprehensive the services to Man provided by the natural environment really are:

- *Photosynthesis.* All human food is derived either directly or indirectly (via the food chain through one or more lower animals) from photosynthetic plants, either on land or in the sea. These plants utilize carbon dioxide and water extracted from the air or earth, with the help of sunlight, to build cellulose and glucose (food energy). Plants also synthesize amino acid molecules—the basic building blocks of protein and, therefore, of living tissues—from the same ingredients, plus traces of nitrogen, phosphorus, and sulfur. Man cannot yet do these things, even in the most sophisticated laboratory.

- *Water.* All plants and animals require adequate water to carry dissolved gases (oxygen, CO_2) and sugars to remove excess heat and (in the case of animals) metabolic waste products like urea. The evaporation–precipitation cycle that replenishes the water supply on land for agriculture as well as for direct human consumption is also a vital part of the global heat redistribution and balancing system that maintains the temperate zone at a comfortable temperature (and keeps the polar snows and glaciers at bay). The oceans constitute the other major element of the climatic regulatory system. Ocean currents carry excess heat away from the equatorial regions and moderate the temperature of coastal areas of the northern land masses.

- *Oxygen and carbon dioxide cycle.* As already mentioned, plants require CO_2 (and sunlight) to manufacture cellulose, glucose, and protein. They produce oxygen as a waste. Oxygen, in turn, is required by all animals, including Man for purposes of energy-metabolism, which yields CO_2 as a waste. The O_2–CO_2 cycle is not necessarily in balance, however, since CO_2 is also being produced in large quantities by the combustion of fossil fuels—accumulated by photosynthetic processes long ago. Some, but not all, of the excess CO_2 is absorbed in the oceans.[†] The remainder remains in the atmosphere, where the concentration has been apparently rising steadily for as long as accurate measurements are available.[35] Increasing amounts of CO_2 in the atmosphere, in turn, can cause a net warming of the surface of the Earth (other environmental factors remaining equal). This is the so-called "greenhouse effect."

- *Nitrogen cycle.* Nitrogen is a minor but essential component of living matter that is also "cycled" by nature. Higher plants and animals are incapable of utilizing gaseous nitrogen as it occurs in the atmosphere. Elemental nitrogen can be "fixed" in small quantities by some natural geophysical processes—such as atmospheric electrical discharges ("lightning")—and by highly energy-intensive industrial processes (ammonia synthesis).

But most nitrogen fixation is due to specialized bacteria working in association with certain plants (legumes).

- *Decay.* As already noted, accumulation of living matter or biomass begins with photosynthesis. The photosynthetic plants support a number of layers, or trophic levels, of increasingly specialized non-photosynthetic animals, culminating (in some sense) with the "top" predator of all, *Homo sapiens*. But the flow of organic material up the trophic ladder must be balanced by a reverse flow. In this process, the bodies of dead animals (and plants) are consumed and utilized. This function is accomplished by a host of successively smaller scavengers from crows and rats, to lobsters and crabs, oysters, insects, worms, fungi, and ultimately bacteria. By means of this network, virtually all of the available energy content of the organic material is used up and converted into entropy.[†] The materials themselves are returned to inorganic forms, when the process is complete.

- *Stability.* The various biological cycles are maintained through a complex web of interrelationships between species. Energy and materials are transferred from organism to organism mainly by ingestion. But relationships between predators, parasites, and prey are inherently quite unstable: Population "explosions" are regularly followed by "busts" in simple ecosystems where only two or three interacting species are present. Nevertheless, stability is a particularly important environmental attribute to Man because of his extended childhood and adolescence and (comparatively) low reproductive rate. The high degree of stability exhibited by temperate and tropical ecosystems is due to the very large number of different species that are interacting. Thus Man has a definite stake in the continued existence and viability of many species of plants and animals—especially fellow predators—even though some of these species may compete directly with Man for the same sources of food.

- *Phylogenetic information.* The process of evolution may be viewed as a continuing large-scale biotic experiment that has extended over several hundreds of millions of years. The "purpose" of the experiment cannot be ascertained, but a great

---

[†]The best available estimate is $42 \pm 7\%$. See Broecker et al.[34]

[†]Entropy is, essentially, permanently unavailable energy. (Some "waste energy" can, in fact, be utilized, but only to the extent that it is *not* unavailable in the thermodynamic sense.) This topic is discussed at some length in Chapter 3.

many potentially useful and valuable bits of information have been accumulated in the form of self-reproducing genes. Genetic structure determines both the morphology and the instinctual behavior patterns, or "programming", of each species. Each species is a suboptimal solution to a survival problem set by nature. (The failed experiments were eliminated along the way by more successful competitors.) This "solution" often involves unique metabolic adaptations, such as the ability to synthesize certain amino acids or vitamins from crude materials, or the ability to digest highly stable materials such as cellulose, or to neutralize certain toxins, or to sense and interpret signals in various parts of the electromagnetic or acoustic spectra. Every time an established species disappears, potentially valuable information is lost in a very literal sense. As yet, Man does not put a very high value on this vast storehouse of genetic data because he has only just begun to learn how to utilize it systematically.

The development of disease-resistant hybrid varieties of corn and wheat, beginning in the nineteenth century, was perhaps the first deliberate use of this untapped phylogenetic library. The discovery and exploitation of antibiotics for controlling the major infectious bacterial diseases was another. At present, scientists are actively searching for biological controls for pathogens of crop plants, to replace broad-spectrum chemical pesticides that are increasingly ineffective because of a rapid buildup of natural immunity in the target pest populations. It is hard to say what other benefits may be obtained in the future by learning how to read the material in this natural library. Certainly, it would be of great value to Man to learn how to duplicate on an industrial scale the key biological processes of photosynthesis and nitrogen fixation. And, it may be important, too, to exploit the abilities of some micro-organisms to extract and concentrate toxic or radioactive materials (like mercury or strontium 90) from highly dilute streams. But one cannot know in advance what knowledge gained from nature will turn out to be useful for Man's purposes. One can only point out that the continued survival of as many different species of organisms as possible secures a potential source of information—and consequently a valuable service —for Man.

• *Public Service Functions.* In addition to the foregoing one might lump together a number of other "housekeeping" functions. In the words of Charles A. S. Hall[36]:

These processes may in some cases be duplicated by man's industrial machinery although this is generally difficult, expensive and incomplete. A number of examples would include: storing of water, maintaining soil structrue and productivity, maintaining the chemical equilibrium of air and water, maintaining an epidemic-free environment through natural predators and diverse plant communities, moderation of microclimates where natural vegetation is profuse, natural fertilization of agricultural land on undammed floodplains, and the cleansing of soil, air and water via natural biological metabolism.

The major point of the foregoing section—which is not a complete catalog of economic benefits derived from the natural world, by any means—is *not* simply that Man obtains valuable services, as well as disservices, from the environment without monetary payment. Nor is it my major point that these nonmarket phenomena, or externalities, result in distortions in the allocative function of the market— deviations from the ideal state of Pareto optimality. The issue is that classical economic theory assumes that the perturbing influence of the externalities is necessarily small and that they can be dealt with within the traditional paradigm by invoking the postulate of *ceteris paribus*. To quote from E. J. Mishan[37]:

The form in which external effects have been presented in the literature is that of partial equilibrium analysis; a situation in which a single industry produces an equilibrium output, usually under conditions of perfect competition, some form of intervention being required in order to induce the industry to produce an "ideal" or "optimal" output. If the point is not made explicitly, it is tacitly understood that unless the rest of the economy remains organized in conformity with optimum conditions, one runs smack into Second Best problems.

"Second Best" is the term used in the literature of theoretical economics to characterize a suboptimal allocation of resources, resulting from a nonmarket condition that cannot be corrected for by any feasible government (or other) intervention.[38] Yet the unmistakable implication remains that a partial equilibrium analysis is a reasonable approximation to a general equilibrium solution and that Second Best is reasonably close to First Best.

But the second and possibly more far-reaching consequence of the pervasiveness and magnitude of social and environmental services that are not—and can never be—allocated entirely by a market price mechanism (or some surrogate of a market) is this: *The basic paradigm of classical economics is too narrowly limited and must be revised and extended.* The nature of the needed changes are discussed at greater length in Chapter 3.

1.4 THE NEED FOR FURTHER DISAGGREGATION

A key characteristic of the "new" problems in economics is that economists are increasingly concerned with the micro-scale. It is important to examine the individual trees, as well as the whole forest. Resource and environmental concerns tend to involve consideration of technological and geographical particulars, as contrasted with interrelationships among broad aggregates such as population, labor force, capital formation, productivity, consumer price indexes, and the like. Life for economists would be easier, if the familiar "quarterly indicators" were always sufficient. But analyses involving comparisons of specific technologies or locations quickly lead beyond the bounds of the traditional curriculum. Thus economists are increasingly being called upon to answer questions to which the simple models and idealized solutions provide no real assistance.

In order to find answers to many of the pressing problems of an era of rapid technological change, an economist must be able to carry out his analyses at a level of aggregation that takes into account the widely different *resources, materials, forms of energy, and production processes* to which technological changes specifically apply.

All this is not in any sense a denial of the value of broader aggregates and indicators in macro-economic analysis. They will always be needed for a wide variety of purposes. The more detailed analyses supplement rather than replace the studies based on aggregate indicators.

A list of examples of "new" problems facing economists arising from the resource–environment–technology interface exemplifies the need for technical detail. Almost every major technological decision has a resource–environmental dimension. The controversy over the SST is an excellent illustration. To build and operate such a plane will result in increased demands for hydrocarbon fuels, and will also result in physical and chemical disturbances to the stratosphere that may ultimately affect the intensity of both ultraviolet and visible solar radiation on the Earth's surface. All of these impacts have potential economic consequences of significant magnitude that require assessment.[39] Similarly, problems arising initially out of perceived resource needs immediately reveal technological and environmental aspects. Thus the exploitation of Alaska's north slope involved building an enormous north–south pipeline across Alaska, with immense potential for environmental disturbance. Proposals to exploit Colorado oil shale or Wyoming–Montana coal are seen to require diversion of large amounts of scarce water away from traditional agricultural uses. The proposals to solve a resource problem by relying more heavily on nuclear power, especially "breeder reactors" and plutonium reprocessing, are also evidently fraught with environmental risks.

The history of the last few years has been—and undoubtedly of the next several decades will be—increasingly preoccupied by the need to choose among and between complex, expensive, and uncertain technological programs, each of which involves large potential risks and hazards as well as possible benefits. The cheap alternatives and easy choices are no longer available.

The economic factors that must be weighted to arrive at rational choices are intrinsically concerned with detailed technological and environmental questions. What is the marginal impact on agricultural output of Wyoming from adding (or subtracting) one unit of water? What is the marginal impact on electricity costs attributable to implementing an adequate nuclear safeguard system? How will these costs affect other sectors? What are the economies of scale for high temperature gas-cooled nuclear reactors, vis-à-vis boiling water reactors? How much will hydrogen cost, via large-scale electrolysis of water by 1990? How much will it cost to desulfurize coal? What are the potential markets for by-product sulfur?

1.5 SOME CRITERIA FOR MODEL-DESIGNERS

Lawrence Klein has defined a *model* as a "schematic simplification that strips away the non-essential aspects to reveal the inner workings, shape, or design of a more complicated mechanism."[40] The fact that a model cannot hope to reproduce "all" the details, even the more important ones, of a nation's economy is an inherent limitation of what can be accomplished by it. Yet its comparative simplicity is also a strength. An excessively detailed model would be cumbersome to handle and expensive to maintain, yet still imperfect. Moreover, having too many parameters with hidden mutual dependences can lead to biased estimates and forecasts with unnecessarily large variances. On the other hand, the simplification and abstraction implicit in a "parsimonious" model can be a trap; for if the model omits key factors that may have a determining effect on possible outcomes, it can depart too far from reality and lead to false conclusions. It is all too common for even experienced investigators (who should know better) to be hypnotized by the neat rows of figures emerging from a computer and mistake them for a portrayal of the real world.

Qualitative versus Quantitative

The first and most important distinction that needs to be made is between qualitative and quantitative models. The term "qualitative" may seem inappropriate as applied to models, but it is not. Diagrams and pictures—said to be "worth a thousand words"—are clearly models, by Klein's definition. (Even a photograph is a "model," since it represents three-dimensional dynamic reality in a two-dimensional static form.) For that matter, a verbal description of a real event is also a kind of model. Qualitative models can be highly precise and rigorous in expressing certain types of information. A classic example is the famous periodic table of the elements, developed by Mendeleev. It is clearly a model, and clearly qualitative. There are no measures involved. However, despite the obvious importance of qualitative models, the remainder of this book is concerned explicitly with quantitative models.[†]

Explanation versus Optimization

A second fundamental dichotomy can be drawn between simulation or forecasting models and optimization models. While both types are applicable in

---

[†]This section is largely taken from a previously published paper by the author.[41]

economics, the latter type is more fundamental to the discipline, whereas physical models are more likely to be of the former kind. Indeed, economics has generally been defined as the study of optimal allocation of scarce (limited in availability) resources among competing ends or uses. In ordinary language, this process is often simply called "economizing."

In particular, classical economics is concerned with the allocation of investment capital among competing projects and localities, the allocation of income among competing expenditure categories (or means of achieving satisfaction), and so on. The newer problems of economics, as already noted, concern optimal allocation of natural resources among sectors of society, selection of technologies to maximize desired outputs and minimize costly inputs and/or costly wastes, and selection of optimal pollution abatement strategies and investment schedules.

Static versus Dynamic

A third key distinction is between static and dynamic models. The static–dynamic distinction concerns the treatment of time. In a static model, time is not a variable: the solution is valid either for a particular point in time (only), or for *all* time, depending on whether the exogenous variables of the problem are themselves time dependent or not. "Cross-sectional" survey data is often used. In a dynamic model, time is an explicit variable, and the solution evolves with time. Longitudinal (time-series) data is appropriate. Clearly, the dynamic optimizing case is most general, but it is also most difficult to formulate (and to compute). Static nonoptimizing models are not particularly commonplace. The only significant example I know of in economics is the input–output model. Combining the two dimensions, there are four major categories:

- Static explanatory models,
- Dynamic simulation or forecasting models,
- Static optimizing models,
- Dynamic optimizing models.

Causation versus Correlation

Quantitative models may also be divided along another axis, depending on the use of phenomenological causality. This is not a distinction that can always be made clearly by an outside observer, since it sometimes depends on the modeler's intentions. There is a wide spectrum. At one extreme, we find "blind" extrapolation of exponential curves as straight lines on log-paper—where no strict causality is even suggested and the forecaster assumes (in effect) that the curve has a kind of independent existence. At the other extreme is the completely causal model where everything is explained in terms of a fundamental physical theory, such as the "laws of gravitation." Obviously, even fundamental theories are not unchangeable. (Too many have been upset during the present century for us to feel confident that we know all the basic laws governing matter and energy, still less living organisms.) Thus dependence on "causality" is always relative.†

Econometric models, based on correlative relationships between variables determined by statistical analysis of time-series data, tend to be weak in causality. In a sense, the use of sophisticated statistical methods can be a substitute for understanding underlying mechanisms and relationships. (This is not necessarily the case: A good theoretical model can be improved by the application of refined statistical techniques for empirical estimation of key parameters. But often statistical means are employed to select relationships between variables precisely because fundamental theory is lacking; indeed, the theory may develop more rapidly as a consequence of such analysis.) Causal models may be either deterministic or probabilistic in structure.

Realistic versus Abstract

Causal models, based on fundamental theory—as opposed to correlative models developed by statistical analysis of empirical data—may be either realistic or abstract. In the latter case, they are intended to represent real phenomena and to assist either in forecasting or in determining optimum arrangements. On the other hand, abstract (data-free) models are intended only to generate "theorems," elucidate limiting cases, and so forth. They tend to be deliberately oversimplified. Much of theoretical resource environmental economics in recent years has been concerned with exploring the properties of very

simple models that (presumably) exemplify broad general principles.

Short-term versus Long-term

An important subsidiary distinction, applicable to realistic forecasts only, must be made between *short-term* and *long-term* dynamic deterministic models. The distinction is extremely important in practice, yet often overlooked (or not understood) even by many practitioners.† The difference hinges on how data is used in the model and how the results are interpreted. Briefly, long-term models are concerned with moving averages or trends, in which temporary departures from equilibrium are deliberately ignored. Indeed, the more precisely calibrated the model is for use in short-term forecasts, the faster it will depart from the long-term trends. Conversely, the more closely it is tied to long-term trends, the less accurately it will reproduce short-term fluctuations. Hence fluctuations in the input (i.e., historical) data are regarded as "noise" and are normally smoothed over. On the other hand, short-term models are specifically concerned with the fluctuations away from equilibrium. Hence they tend to utilize *all* historical data, with the smallest possible time increments—usually quarterly—and, as a rule, recalibrate and recompute after each new set of data points are added to the series. And because the nonequilibrium aspect is vital, multiple regression analysis is heavily used in developing the forecasts.

The important point here is that short-term models are *predictions* because they take off from an actual set of initial conditions, such that all values of variables and parameters are guaranteed to be realistic, at least, at time zero. The predictive value declines fairly rapidly as the forecast horizon is extended, of course, because the starting point is always off-equilibrium.

Long-term models, on the other hand, do not give predictions, but are used instead to make *projections* usually in sets of possible alternatives. A long-term model has no predictive value even in the short-run, because it is concerned with trends and smoothed

---

†A causal model may also yield predictions in terms of probabilities or "expectation values," as in decision models.

†To be as careful as possible, short or long should be measured in terms of the unit of time, or the interval, that is appropriate to the problem. Short-term models typically use quarters as steps. Long-term models would normally measure time in 5 to 10 year periods.

averages. These are seldom in agreement with actual current values of the model variables. And because equilibrium conditions are mainly of concern, it is reasonable to depend heavily on accounting identities (e.g., input–output relationships, materials, and energy balances) that can be relied upon to change fairly slowly, if at all. The purpose of a long-term model is to examine the quantitative consequences of changes in exogenous trends in parametric relationships or in constraints. Conclusions drawn from long-term models should always be explicitly *contingent* on the particular set of starting assumptions. (A contingent statement is always of the type: "if . . . then . . . ." The assumptions are an intrinsic part of the statement.)

Because short-term and long-term models are used for different purposes (and supported by different institutional sponsors), the developers are often out of touch with each other and, at times, unjustifiably contemptuous of each other's methodologies. In particular, there is a tendency for each to encroach on the others' domain, rather than to develop a synergistic dialog. But it is becoming increasingly urgent that such a dialog be initiated, both to minimize duplication of effort and rediscovery of the wheel and for more fundamental reasons. The latter have to do with the need to develop realistic long-term dynamic optimizations in which short-run departures from equilibrium play an explicit role.

Levels of Aggregation

There exists a natural hierarchy of aggregation levels for models in economics, each level being useful for some particular purpose. Most highly aggregated (call this Level 1) are the models using such broad measures as total population, total labor force, unemployment rate, Gross National Product, gross capital formation, private consumption expenditures, wholesale and consumers' price indexes, and rate of price inflation. Noneconomic models at the same level would introduce total energy flow, biomass, and the like. Some of these aggregates, notably population and labor force, can be estimated for a number of years ahead because birth and death rates change relatively slowly, and age distributions of the population can be projected with fair accuracy.

Beyond these pivotal aggregates, forecasts of economic quantities become increasingly difficult and have to be based on a priori assumptions, such as the proportionality of inputs to outputs or the

assumption that relationships between the various quantities that have held true during past periods of time will continue to hold true in the future. Such relationships often take the form of equations, by which, if certain macro-variables such as population, labor force, and productivity are assumed, the other quantities can be estimated. The trends projected by such techniques are more reliable for short-term forecasts than for long-term ones; and, therefore, as forecasts extend further and further into the future, they are likely to become increasingly unrealistic.

A special problem with long-term forecasts at this level of aggregation is that the key elements affecting long-term changes in the economy are not necessarily the same as those affecting the short-term. For the long-term, the key elements include technological changes and possibly raw material shortages, rising energy costs, material substitutions, changes in social customs, changes in the educational level of the population, environmental deterioration, and many other factors not explicitly accounted for in the aggregates generally used in Level 1.

A second level of aggregation (call it Level 2) involves dividing up the economy by *industry* sector and/or *region*. A familiar example of this is the input–output model that takes the form of a matrix recording the pattern of flow of materials and energy (or the pattern of purchases and sales) between industry sectors, between each sector and the government, and between each sector and the final customer. Such a table does not identify the particular commodities or energy forms that flow into and out of the sectors, nor the transformations that take place in the production processes, but it accounts for all inflows and outflows *in total*. These inflows and outflows must balance, both for the system as a whole and for each individual sector after accounting for waste and for materials drawn from the environment.

The development of input–output models provided for the first time a comprehensive view of the *structure* of the economy, like a still photograph that catches an action in mid-motion. These models facilitated undertaking studies that had formerly been extremely tedious, if not impossible. With them, one could determine the effects a change in one sector might be expected to have on the other sectors. For example, if automobile size and weight are reduced, the direct effects on the iron and steel, coal mining, petroleum refining, glass, nonferrous metals, synthetic rubber, chemical, machine tools, and many other sectors can be traced. But beyond these direct effects are secondary and tertiary effects: The reduced demand for intermediate products entering into the automobile industry affects communications, transportation, electric power, and so on. The effects ripple through a maze of interrelationships.

For some industries, the level of aggregation used in published input–output tables, even if regionalized, is still much too broad for accurate analysis. For example, the sector "Industrial Chemicals" includes a wide variety of products made from different raw materials by different processes and used in different sectors. Thus benzene is derived both from coal and from petroleum. It is used in petroleum refining (to gasoline), in manufacturing synthetic rubber, in making the plastic styrene, and in many other chemicals. A shift in the proportions of these products within the sector can alter significantly the inputs and outputs for the sector.

At the commodity level of disaggregation, it is apparent that the same commodity can, in many cases, be produced by several alternative processes. An example is the production of polyvinyl chloride (PVC) bottles. At least 20 different processes may be involved in the manufacture of PVC or its intermediates and these can be combined in more than 60 different ways—each with different environmental impacts. In other words, there are over 60 possible *chains of processes* leading from raw materials to finished PVC, all requiring different inputs and yielding different wastes. A study of the environmental consequences of regulations affecting process technology or energy would need to take these alternative chains of processes into account. To deal with problems where this level of detail is unavoidable requires a still higher lever of disaggregation (Level 3).

Determinism versus Uncertainty

A final distinction of utmost importance must be made between models that are endogenously deterministic with fixed inputs (in the "clockwork" sense) and models that explicitly provide for probabilistic irregularly variable or random inputs. Variable inputs may be distributed according to different rules, ranging from normal (Gaussian) or log-normal distributions to "Poisson events," such as floods, fires, epidemics, earthquakes, and so on, to ad hoc heuristic "scenarios" *based on extra-model intelligence.*

For certain types of physical models, stochastic or normal distributions are common. For instance, the "mix" of local weather conditions, genetic variability, or other factors is likely to be subject to a normal type of distribution. The same thing may be true of certain behavioral variables, such as preference functions or "willingness to pay." Or the measurement process may be subject to random influences of various sorts that introduce errors that are distributed in a normal fashion.

There are many situations, especially in the sociopolitical realm, where it cannot be assumed that the measurement process or the phenomena being measured behave in any such regular fashion. In particular, it is often the case that any attempt to measure a factor will cause it to change. This is analogous to the "uncertainty principle" of quantum physics, which states that certain combinations of variables can never be known precisely at the same time, since a measurement of one will alter the other. A similar indeterminacy principle enters the picture if one attempts to predict the behavior of individuals, committees, or structured organizations (including governments) having a small number of effective decision-makers. This is a logical implication of the "free will" of the individual. But, even if humans were actually instinctually preprogrammed, in the same sense as insects, an indeterminacy principle would still be applicable because it is clearly impossible to monitor an individual human's behavior (including thoughts) closely enough to predict his or her actions without disturbing the object of the surveillance to the point of affecting those actions significantly.

To deal with indeterminacy in the realm of human behavior, it is convenient to introduce "policy" variables (or parameters) in the models. Rather than trying to predict what human behavior will be, which is impossible where a small number of effective decision-makers are involved, the model must be formulated to explore the implications of alternative decisions or "policy options."

Mapping Model Types to Issues

How can one decide what type of model is most appropriate to a given policy issue? The foregoing taxonomy constitutes a kind of hierarchical checklist for the classification of *models*. In effect, it defines a large number of possible "pigeon holes." On reflection, it is clear that the same screening process can also be applied to *problems* by examining the relevant variables and relationships:

- Are the pertinent data quantitative?
- Is the question one of prediction? Or is it a question of characterizing the best (optimal) solution?
- Is time an explicit variable?
- Is there an underlying phenomenological theory available? Or must one rely on observed correlations between independent observations?
- Is realism desired? Or is it the object of the exercise to deepen our understanding of the fundamentals by systematic simplification?
- If time is a variable and a realistic deterministic forecast is desired, is the time scale short or long?
- What is the level of aggregation at which the key phenomena are observable (viz., national, sectoral, regional, commodity, process)?
- Are there intrinsic uncertainties or random elements in the problem? Is the behavioral response by individual decision-makers or small groups of decision-makers a factor in the problem?

Disregarding the last three items for quantitative models there are 12 possible combinations of these characteristics, of which eight or nine (at least) seem to be relevant (i.e., the boxes are occupied). For two of the categories, a further short–long subdivision seems called for. Any of the cases may be at any level of aggregation and may involve random influences or unpredictable human intervention. As noted, classifications are not always unambiguous. For instance, any "equilibrium" air pollution dispersion (e.g., smoke "plume") model would certainly be classed as a causal–empirical, but it is not quite clear whether the term "static" or "dynamic" is applicable unless the model is more precisely identified (there are many possible plume models). A pollution-forecasting model utilizing empirically determined pollution coefficients for highly aggregated sectors would certainly be classed as static simulation, but there might be a question as to whether truly causal relationships are used. If the pollution output for a sector were developed based on more detailed process-level analysis, explicitly incorporating materials and energy balances, there would be less ambiguity, of course.

Central to the model design problem is the decision on what factors should be explicitly taken into

	Realistic		Nonrealistic
	Noncausal (Statistical)	Causal–Empirical	Causal–Abstract
Static Explanation	?	X	X
Dynamic Forecasting	Short Long	Short Long	X
Static Optimization		X	X
Dynamic Optimization			X

account. This judgment is determined by the questions the model is designed to answer. If the investigator wants to know the general trend of world food supply vis-à-vis demand over the next decade, he may simply project world population and trends in Calorie output per cultivated acre and compare the two. This can be useful, for some purposes, although it does not provide much information regarding costs, prices, incomes, trade, or other important matters. The general conclusions, of course, hang on the accuracy of the projections of the two aggregates: population and Calories. If, however, the question is What kinds of food can best be produced in particular climatic zones, in the face of specific environmental hazards, under given labor and energy costs and market prices? a more detailed analysis—by country and by type of product—is needed. If the question is, further, What contribution can technology make to increase food production? How much new capital will be required? Where will the necessary capital be obtained? and What social and political changes will be needed to implement the innovations? a much more detailed analysis still is needed.

To accomplish the practical purposes of applied economics, finally, we must begin to build appropriate computerized accounting and optimization models to reflect these factors. No doubt, such models will in many cases be direct extensions of, or analogous to, the business cycle and input–output models that were developed in response to the economic management needs of the 1930s. There are some respects, however, in which the models that must be built will be of a different kind, because they reflect transformations and flows of fundamentally different character than the monetary flows, around which the earlier models were designed. Perhaps equally relevant, the capacity of electronic computers to store and manipulate enormous quantities of data has grown by many orders of magnitude in the last two decades alone. Thus some new modeling concepts and mathematical techniques are introduced to exploit these capabilities. Last, but certainly not least, one must address the problem of developing an appropriate statistical data base for this new generation of computer models.

Corresponding to the foregoing division, the next section of the book (Chapters 2 and 3) deals with basic economic theory as related to material and energy flow in the economy, and externalities associated therewith. The second section (Chapters 4 and 5) is concerned with the development of appropriate models to simulate these flows, and to optimize national resource allocation and choice of technology, location, investment strategy, pollution abatement strategy, regulatory strategy, and so forth, taking into account the real constraints associated with resource distribution, production capacity, technology, and the like. All of this has to be expanded upon. In the third section (Chapter 6), I discuss the requirements for data and the organization of an appropriate statistical system on an international basis.

REFERENCES

1. H. L. Moore, *Economic Cycles: Their Law and Cause*, Macmillan, New York, 1914.

2. W. C. Mitchell, *Business Cycle: The Problem and Its Setting*, National Bureau of Economic Research, New York, 1927.

3. J. Schumpeter, *Business Cycles*, Vols. I and II, McGraw-Hill, New York, 1939.

4. J. M. Keynes, *The Means to Prosperity*, Harcourt and Brace, New York, 1933.

5. J. M. Keynes, *The General Theory of Employment, Interest and Money*, Macmillan, London, 1936.

6. J. Tinbergen, *An Econometric Approach to Business Cycle Problems*, Herman et Cie, Paris, 1937.

7. J. Tinbergen, "Statistical Testing of Business Cycle Theories I, II," League of Nations, Geneva, 1939.

8. L. R. Klein and A. S. Goldberger, *An Econometric Model of the United States, 1929–1952*, North-Holland, Amsterdam, 1952.

9. F. Quesnay, *Tableau Economique*, published 1758, Bergman, New York, 1968.

10. W. Leontief, "Foundations of Soviet Strategy for Economic Growth," in N. Spulber (ed.), *Selected Soviet Essays, 1924–1929*.

11. W. W. Leontief, "Interrelation of Prices, Output, Savings and Investment: A Study in Empirical Application of Economic Theory of Interdependence," *Rev. Econ. Stat.*, 19 (August 1937).

12. W. W. Leontief, *Structure of the American Economy, 1919–1939*, 2nd ed., Oxford University Press, New York, 1951.

13. S. Kuznets, *National Income and Its Composition, 1919–1938*, New York, 1941.

14. S. Kuznets, *National Income: A Summary of Findings*, New York, 1946.

15. R. Stone, "Definition and Measurement of the National Income and Related Tables," Memo to Subcommittee on National Income Statistics, Committee of Statistical Experts, League of Nations, Princeton, NJ, June 1946.

16. United Nations, *System of National Accounts*, Studies in Methods, Series F, No. 2, Rev. 3, New York, 1968.

17. M. K. Evans and L. R. Klein, "The Wharton Econometric Forecasting Model," Studies in Quantitative Economics No. 2, Economic Research Unit, 1968; Also M. D. McCarthy, "The Wharton Quarterly Econometric Forecasting Model: Mark III," Studies in Quantitative Economics No. 6, Economic Research Unit, 1972.

18. J. S. Duesenberry, G. Fromm, L. R. Klein, and E. Kuh (eds.), *The Brookings Quarterly Econometric Model of the United States*, Rand McNally, Chicago, 1965.

19. F. deLeeuw and E. Gramlich, "The Federal Reserve–MIT Econometric Model," *Fed. Rev. Bull.* (January 1968). Also R. H. Rasche and H. T. Shapiro, "The FRB–MIT Econometric Model: Its Special Features," *Amer. Econ. Rev. Papers Proc.*, 58 (May 1968).

20. S. H. Hymans and H. T. Shapiro, "The Michigan Quarterly Econometric Model of the U. S. Economy," Research Seminar on Quantitative Economics, University of Michigan, 1973.

21. L. Thurow, "A Fiscal Policy Model of the United States," *Sur. Cur. Business*, 49 (June 1969).

22. M. K. Evans, *Macroeconomic Activity, Theory, Forecasting and Control*, Harper and Row, New York, 1969.

23. R. J. Ball (ed.), *The International Linkage of National Economic Models*, North-Holland, Amsterdam, 1973.

24. C. Almon, Jr., M. Buckler, L. M. Horwitz, and T. C. Reimbold, *1985: Interindustry Forecasts of the American Economy*, Heath, Lexington, MA, 1974.

25. U. S. Bureau of Labor Statistics, *The Structure of the U. S. Economy in 1980 and 1985*, Bulletin 1981, USGPO, Washington, DC, 1975.

26. W. W. Leontief, A. Carter, P. Petri, and J. Stern, "Technical Report on the Study of the Impact of Prospective Environmental Issues and Policies On the International Development Strategy," UN Center for Development Planning Projections and Policy Office, 1976.

27. I. Fisher, *Nature of Capital and Income*, A. M. Kelley, New York, 1906.

28. F. Knight, *Risk, Uncertainty and Profit*, A. M. Kelley, New York, 1921.

29. K. Lancaster, *Consumer Demand: A New Approach*, Columbia University Press, New York, 1971.

30. A. C. Pigou, *Economics of Welfare*, 3rd ed., Macmillan, New York, 1938.

31. Paul Samuelson, "Pure Theory of Public Goods," *Rev. Econ. Stat.* (1954).

32. J. E. Meade, "External Economics and Diseconomies in a Competitive Situation," *Econ. J.*, **62** (1952).

33. B. Commoner (ed.), *Energy and Human Welfare: A Critical Analysis*, Macmillan, New York, 1975.

34. W. S. Broecker, Y. H. Li, and T. H. Peng, Chapter II, in D. W. Hood (ed.), *Impingement of Man on the Oceans*, Wiley-Interscience, New York, 1971.

35. C. A. S. Hall, C. A. Ekdahl, and D. E. Wartenburg, "A Fifteen Year Record of Biotic Metabolism in the Northern Hemisphere," *Nature*, **255** (May 8, 1975).

36. C. A. S. Hall, "The Biosphere, the Industriosphere and Their Interactions," *Bull. Atom. Sci.*, **XXXI** (March 1975).

37. E. J. Mishan, "Reflections on Recent Developments in the Concept of External Effects," *Canad. J. Econ. Polit. Sci.*, **31** (February 1965).

38. O. Davis and A. Whinston, "Welfare Economics and the Theory of the Second Best," *Rev. Econ. Stud.*, **32** (January 1965).

39. U. S. Department of Transportation, *Economic and Social Measures of Biologic and Climatic Change*, CIAP Monograph 6, Washington DC, September 1975.

40. L. Klein, unpublished manuscript.

41. R. U. Ayres, "A Taxonomy of Environmental Models," in W. R. Ott (ed.), *Proceedings of Conference on Environmental Modeling and Simulation*, sponsored by EPA, April 19–22, 1976.

CHAPTER TWO *Dynamic Optimizing Models of Resources and the Environment*

Whereas the previous chapter provided background as to the need for, and use of, models of the flow of materials and energy through various transformation stages in the economy, the present chapter reviews some relevant antecedent material in the literature of economic theory.

Realistic quantitative large-scale resource-environmental models are, of course, discussed in Chapters 4 and 5. However, the mainstream literature of economics seldom refers to—or even recognizes the existence of—these large and complex models.[†] Despite (or perhaps because of) formidable mathematical difficulties, economists have evinced a strong preference for ultrasimplified models involving only a few sectors, variables, and relationships, such that the formal conditions for an optimal solution over time can be displayed in the form of differential equations.

It is even customary to allow most of the assumed relationships to remain implicit, being specified only to the extent that the signs of certain derivatives are constrained to remain positive or negative. Numerical solutions are scorned (they would be meaningless anyhow) and the results—or theorems—that are derived are of the most general and qualitative sort.

†Indeed, mainstream economics barely recognizes input–output models—but that is another topic!

There is little discussion in this chapter of simple *static* models; for example, of the "stationary state" of classical economics. By general agreement, such models are of little value to the understanding of processes of accumulation and growth (or of depreciation and decay). The fact that the large-scale models discussed later are not—at present—truly dynamic but, more precisely, "quasistatic" does not affect the statement in any way.

It is not my intention to survey these simple optimizing models in detail. A broad brush treatment seems quite sufficient in view of the very slight relevance any of them have to problems of concern in the real world. However, for the benefit of those readers who are not professionally involved in this type of modeling I have presented some of the basic mathematics in a simplified and nonrigorous form.

The type of model discussed in this chapter is often associated with the name of Frank Ramsey.[1] Ramsey undertook to address the optimal rate of savings over a period of time, subject to some restrictive (and unrealistic) assumptions that facilitated an explicit mathematical treatment using the calculus of variations. Ramsey's model did not throw a great deal of new light on the question at issue—at best it confirmed the obvious by more rigorous reasoning—but it was highly influential among theoretical economists because it considered, essentially for the first time, a nonequilibrium transitional state of affairs, rather than the "ultimate" stationary state that had dominated classical economic thinking on the subject. Harold Hotelling used similar mathematics in an early analysis of optimal depreciation rates.[2]

Both of these studies were attempts to mathematicize the theory of capital accumulation and growth, which was a major intellectual issue to economists in the 1920s. As it happens a number of problems in resource–environmental economics are now perceived to be aspects of capital theory. Hotelling was one of the first to think of the extraction of nonrenewable resources (e.g., mineral ores) as depreciation of a capital stock. In recent decades, the physical and biological environment as a whole has been increasingly viewed as a form of natural capital.

Indeed, the recent convergence of interests between resource–environmental economists and capital theorists can perhaps be understood best in terms of Kenneth Boulding's famous analogy between the open ("cowboy") economy of the past and the closed ("spaceship") economy of the future.[3]

The difference between the two types of economy becomes most apparent in the attitude towards consumption. In the cowboy economy, consumption is regarded as a good thing and production likewise; and the success of the economy is measured by the amount of the throughput from the "factors of production," a part of which, at any rate, is extracted from the reservoirs of raw materials and noneconomic objects, and another part of which is output into the reservoirs of pollution. If there are infinite reservoirs from which material can be obtained and into which effluvia can be deposited, then the throughput is at least a plausible measure of the success of the economy. The gross national product is a rough measure of this total throughput. It should be possible, however, to distinguish that part of the GNP which is derived from exhaustible and that which is derived from reproducible resources, as well as that part of consumption which represents effluvia and that which represents input into the productive system again. Nobody, as far as I know, has ever attempted to break down the GNP in this way, although it would be an interesting and extremely important exercise, which is unfortunately beyond the scope of this paper.

By contrast, in the spaceman economy, throughput is by no means a desideratum, and is indeed to be regarded as something to be minimized rather than maximized. The essential measure of the success of the economy is not production and consumption at all, but the nature, extent, quality, and complexity of the total capital stock, including in this the state of the human bodies and minds included in the system. In the spaceman economy, what we are primarily concerned with is stock maintenance, and any technological change which results in the maintenance of a given total stock with a lessened throughput (that is, less production and consumption) is clearly a gain. This idea that both production and consumption are bad things rather than good things is very strange to economists, who have been obsessed with the income-flow concepts to the exclusion, almost, of capital-stock concepts.

From the standpoint of mathematical models, at least, capital theory did not progress much beyond the contributions of Ramsey and Hotelling until after World War II when a quantum leap forward occurred in the mathematics of optimizing static coupled systems. This technique of finding an "extremum" solution for a set of linear equations subject to a set of linear inequality constraints—known as "linear programming"—was rapidly extended to the case of nonlinear constraints (nonlinear programming) and, finally, time-dependent constraints (dynamic programming).

The resulting body of mathematics is known today as the "theory of optimal controls," the name perhaps suggesting something of the original impetus for the work—which was space technology. Just as the United States and the Soviet Union pursued separate but parellel space programs, the two countries also separately developed the needed mathematical tools.[4, 5] The applications to economics, in general, and capital theory, in particular, were noted by Bellman[4] but particularly exploited by Arrow et al.[6–8] The brief summary treatment in the following section is adapted from a tutorial paper by Dorfman.[9]

2.1 THE MATHEMATICAL THEORY OF OPTIMAL ACCUMULATION OR DEPRECIATION OVER TIME

As noted already, the underlying mathematics are the same whether one is talking about man-made capital (savings, public investment, etc.) or natural capital (exhaustible resources or "renewable" resources). From time to time, the parallels between concepts in the two fields are pointed out, but in general they are quite obvious.

The necessary postulates and vocabulary are as follows:

1. At some initial period t, there exists an inherited capital stock K_t. At later times $\tau > t$ the capital stock is denoted K_τ.

2. The time rate of change of capital stock at time τ is denoted $\dot{K}_\tau$, where $\dot{K}_\tau$ is a function of the current stock of capital itself, as well as a vector of parameter(s)[†] χ_τ, which also are functions of time and which are subject to exogenous manipulation. These are the *controls* (if one is talking about engineered systems) or *policy variables* (if one is talking about firms or governments). This gives rise to the first basic equation, namely,

$$\dot{K}_\tau = F\{K_\tau, \chi_\tau, \tau\} \qquad (2.1)$$

Notice that the decision-making entity (the "controller") is not at liberty to specify K_τ after the initial time; he can only influence K_τ (within limits) via the control parameters, whose impact is not immediate in any case. It is important to observe that, since $\dot{K}_\tau$ is an explicit function of χ_τ, its indefinite time integral, K_τ, is at least an implicit function of these parameters. K_τ *may* depend explicitly on χ_τ.

3. The existing capital stock, at any time, generates some kind of yield. If the stock is a money investment, the yield will be a dividend or interest payment. If it is thought of in physical terms, the yield will be a value added (by capital) to a raw material flow. If one is talking about natural capital of the renewable type (e.g., agricultural land), the yield is the annual output flow, or harvest, per se. It is convenient to designate the yield of the capital as a generalized utility output U_τ, where U_τ is an explicit function of the capital stock at the time, and also a function of the settings of the controls χ_τ. It may also be an explicit function of time itself. Thus the firm or operating entity may wish to count, at time t, the present (i.e., current) value of all future revenues or service yields, by introducing a discount factor, for example, of the form $\exp(-\mu t)$. Hence U must have the form:

$$U_\tau = U(K_\tau, \chi_\tau, \tau) \qquad (2.2)$$

4. To select an optimal path implies that some "objective function" is maximized. The problem has been set up in such a way that the natural choice of objective function W is, simply, the time integral over U_τ from the initial time ($\tau = t$) to some terminal time ($\tau = T$), where T may be

[†]There may be more than one of these parameters.

extended to infinity if that is desired. The value of the objective function clearly depends on the starting value of the capital stock, which is given, and the entire *time path* of the control variables. The notation $\hat{\chi}$ is introduced to convey this path dependence:

$$W(K_t, \hat{\chi}) = \int_t^T U_\tau(K_\tau, \chi_\tau, \tau)\, d\tau \qquad (2.3)$$

The mathematical problem is to maximize $W(K_t, \hat{\chi})$, subject to equations (2.1) and (2.2) and any terminal or other constraints that may be imposed depending on the problem. For convenience, label the values of W corresponding to an optimal time path V^*, where

$$V^*(K_t) = \max W(K_t, \hat{\chi}) \qquad (2.4)$$

Incidentally, it is worth notice that F in (2.1) must be defined explicitly at the outset in order to obtain a definite numerical value for $V^*(K_t)$, but if one is only interested in qualitative relationships among the variables, it may not be necessary to be specific about the functions in question.

The maximization of W can be carried out formally in several ways. The following procedure is adopted from Dorfman.[9] First, rewrite (2.3) to simplify the notation

$$W(K_t, \hat{\chi}, t) = \int_t^T U_\tau(K_\tau, \chi_\tau, \tau)\, d\tau \qquad (2.3a)$$

and now break the time period (t, T) into two parts; a short initial increment $(t, t + \Delta)$ and a longer subsequent period $(t + \Delta, T)$. Then (2.3a) becomes

$$W(K_t, \hat{\chi}, t) = U_t(K_t, \chi_t, t)\Delta + \int_{t+\Delta}^T U_\tau(K_\tau, \chi_\tau, \tau)\, d\tau$$

$$(2.5)$$

This equation isolates the initial value of the capital K_t and the initial value of the control variable(s) χ_t in a separate term. It is important to note that the starting values of capital stock and control variables in the integral over the remaining time period ($t + \Delta, T$) are now, respectively.

$$K_{t+\Delta} \quad \text{and} \quad \chi_{t+\Delta}$$

Next, observe that

$$W(K_t, \hat{\chi}, t) = U_t(K_t, \chi_t, t)\Delta + W(K_{t+\Delta}, \hat{\chi}, t + \Delta)$$

$$(2.6)$$

Now consider all possible time paths for $\hat{\chi}$—including all possible initial values $\chi_{t+\Delta}$—from time $t + \Delta$ onward, and select the one that maximizes the objective function $W(K_{t+\Delta}, \hat{\chi}, t + \Delta)$. Because the selection of optimal time paths was independent of the initial interval Δ, we cannot yet assert that the left-hand side of (2.6) is absolutely maximized by the procedure. Still it is convenient to write

$$V(K_t, \chi_t, t) = U(K_t, \chi_t, t)\Delta + V^*(K_{t+\Delta}, t + \Delta)$$

$$(2.7)$$

where the second term of the right-hand side is a true maximum. By implication, $\chi_{t+\Delta}$ is that value of the control variable at time $t + \Delta$ that corresponds to this maximum. The question now arises: What must be the value of χ_t to maximize the *first* term on the right-hand side of (2.7) and also satisfy the foregoing requirement on $\chi_{t+\Delta}$? When this value for χ_t is found, then $V(K_t, \chi_t, t)$ will automatically attain its maximum possible value $V^*(K_t, t)$.

The ordinary differential calculus can be used at this point to obtain an equation (or equations) for χ_t. Barring certain kinds of pathological behavior on the part of the functions, the maximum will occur when the partial derivative of the right-hand side of (2.7) vanishes. In the usual case, $V^*(K_{t+\Delta}, t + \Delta)$ is not an explicit function of χ_t (by definition), but it is still an implicit function of χ_t through its dependence on $K_{t+\Delta}$. Thus we have

$$0 = \Delta \frac{\partial}{\partial X_t} U(K_t, \chi_t, t)$$
$$+ \frac{\partial V^*(K_{t+\Delta}, t + \Delta)}{\partial K_{t+\Delta}} \cdot \frac{\partial K_{t+\Delta}}{\partial \chi_t} \qquad (2.8)$$

It is convenient at this point to define a new variable

$$P_K(t) = \frac{\partial V^*(K_t, t)}{\partial K_t} \qquad (2.9)$$

In economic terms, $P_K(t)$ is the marginal value (or "shadow price") of capital at time t. (If the yield from the capital is monetary, value has its ordinary monetary meaning; if the yield is a physical product or a service, the term "marginal value" must be correspondingly interpreted.)

For very small values of the time increment Δ (and appropriately smooth, continuous functions), we can use the so-called perturbation series approximations:

$$K_{t+\Delta} = K_t + \Delta \dot{K}_t + \frac{\Delta^2}{2} \ddot{K}_t + \cdots \qquad (2.10)$$

$$P_K(t + \Delta) = P_K(t) + \Delta \dot{P}_K(t) + \frac{\Delta^2}{2} \ddot{P}_K(t) + \cdots$$

$$(2.11)$$

In the standard case, where $\dot{K}_t$ is an explicit function of χ (but K_t is not), substituting (2.9), (2.10), and (2.11) into (2.8) yields

$$0 = \frac{\partial U}{\partial \chi_t}(K_t, \chi_t, t) + P_K(t) \frac{\partial F}{\partial \chi_t}(K_t, \chi_t, t) \quad (2.12)$$

[In the more general case, where K_t is also a function of X_t, it can be shown that an extremum exists only if (2.12) is true and, in addition,

$$\frac{\partial K_t}{\partial \chi_t} = 0 \qquad (2.13)$$

on the optimal path.] Now assume that χ_t—the optimum initial settings for the control variables—are fixed by (2.12). Then, from (2.7),

$$V^*(K_t, t) = \Delta U(K_t, \chi_t, t) + V^*(K_{t+\Delta}, t + \Delta)$$

$$(2.14)$$

Differentiating both sides with respect to K_t, and substituting (2.9), we obtain

$$P_K(t) = \Delta \frac{\partial U}{\partial K_t} + \frac{\partial V^*}{\partial K_{t+\Delta}} \cdot \frac{\partial K_{t+\Delta}}{\partial K_t}$$

$$= \Delta \frac{\partial U}{\partial K_t} + P_K(t + \Delta) \cdot \frac{\partial}{\partial K_t}\left(K_t + \Delta \dot{K}_t + \cdots\right)$$

$$= \Delta \frac{\partial U}{\partial K_t} + \left(P_K + \Delta \dot{P}_K + \cdots\right)$$

$$\times \left(1 + \Delta \frac{\partial F}{\partial K_t} + \cdots\right)$$

$$= P_K + \Delta\left[\frac{\partial U}{\partial K_t} + \dot{P}_K + P_K \frac{\partial F}{\partial K_t}\right] + \cdots$$

or cancelling and simplifying

$$0 = \dot{P}_K + \frac{\partial U}{\partial K_t} + P_K \frac{\partial F}{\partial K_t} \qquad (2.15)$$

Equations (2.12) and (2.15) together constitute the basic equations of the "maximum principle" of the theory of optimal controls.

It is interesting and sometimes helpful to note that the three basic equations—(2.1), (2.12), and (2.15)—are equivalent to the conditions for maximizing the value for a certain auxiliary function of U, P_K, and F with respect to K_t and χ_t. This auxiliary function is called the Hamiltonian, namely,

$$H = U + P_K F \qquad (2.16)$$

The defining equation (2.1) can be expressed as follows:

$$\frac{\partial H}{\partial P_K} = \dot{K}_t \qquad (2.17)$$

whereas the conditions for the "maximum principle" are given by

$$\frac{\partial H}{\partial \chi_t} = 0 \qquad (2.18)$$

$$\frac{\partial H}{\partial K_t} = -\dot{P}_K \qquad (2.19)$$

it can be verified quickly that (2.18) reduces to (2.12), and (2.19) is equivalent to (2.15).

The Hamiltonian formalism is a great convenience for analysis, since the equations determining an optimal strategy or an optimal solution can be written down immediately as soon as the Hamiltonian function is defined.

Up to this point, I have skimmed, with ethereal lightness, over a critical aspect of the optimization problem, that is, the treatment of "time preference" or intertemporal weighting of relative utilities. Any specific mathematical treatment should also be linked to a particular choice of optimality criterion. A clear but profound discussion of the issues involved—many of which are ethical—is given in a recent book by Talbot Page.[10]

Briefly, the analyst must choose, implicitly if not explicitly, between two fundamentally different postures. They follow:

1. To measure the utility of future generations only in terms of the extent to which it is reflected in the feelings of satisfaction enjoyed by the present generation.
2. To measure the utility of future generations *to themselves.*

The first posture is consistent with the common convention of discounting the future in some fashion. The appropriate choice of discounting rule depends to some extent on whose utility is at issue. For financial transactions, the market rate of interest is the obvious choice. For an economy as a whole, the overall rate of productivity growth is also a natural choice. For an isolated agricultural island, the rate of agricultural productivity is the most relevant. For individuals, the choice is not necessarily simple, however. Individuals may have time preference functions quite different from market interest rates or GNP growth rates. A drug addict will sell his body and soul—for delivery tomorrow—in exchange for the few dollars he needs for a "fix" today. An elderly person may discount the future much more than the average, whereas a young person may be willing to forego gratification for a long time. Religious and cultural factors are deeply involved: The middle-class Calvinist or Puritan makes plans far ahead; the urban ghetto teenager lives from one day to the next.

However all of these differences can be—more or less—accommodated within the notion of a discounted present value for benefits (net of costs) to be received in the future. Page has labeled this the "present value criterion" for utility maximization. It has a number of advantages: most notably that it is *complete* in the sense of being a sufficient prescription for allocating resources, goods, and services over time and that it is economically *efficient.*[†] It is consistent with the results of a (perfected) free market.

It also has some fundamental disadvantages, of which the most important may be that it does not guarantee survival of future generations of humans, still less geomorphic or biological assets that do not reproduce themselves as fast as the prevailing social discount rate. From a strictly theoretical point of view, too, the present value criterion is deficient in another sense: In optimizing only for the benefit of the first generation, it generates a future time path for the allocation of resources that may not (and, in

[†]This is true, at least in the limited (Kaldor) sense, that the "winners" *could* compensate any "losers" and still come out ahead, though no actual compensation may take place.

general, will not) be followed by subsequent generations since it is not optimal for them. This means the present value method is fundamentally inconsistent, since the utility of the first generation is actually only maximized if the later generations follow the originally optimal path. Perhaps a better way of putting it would be to say that a time path can only be truly optimal in the present value sense, if the *same* path will be optimal to each succeeding generation. This self-consistency requirement strongly limits the allowable methods of discounting. (As it happens, the familiar exponential discounting function does have the desired property in some circumstances; however, a more detailed discussion of this topic would carry us too far afield.)

The alternative posture is one where future generations are allowed (in principle) to have an equal vote in the selection of an optimal time path. This implies a zero rate of discount for the future, but it does not specify what exactly should be maximized. In a general sense, the optimal path implied by the criterion requires that each generation should pass on its "resource base" without depletion or impairment. What this means, exactly, is still very much open to dispute. Both the concepts of "resource" and "depletion" mean different things to different people. Page suggests a rule that will appeal to economists: Resources and depletion should not be measured in physical terms, but in terms of their prices, relative to the overall economy. The "no depletion" rule means that the overall index of virgin resource prices should be held constant (or declining). A number of management options exist, in principle, including severance taxes, increased recycling, and technological substitutions, for securing this result.

In the remainder of this chapter the formalism associated with the present value criterion, with a time-preference function $e^{\rho t}$, is used because it is convenient and familiar. The conservationist case corresponds to a zero discount rate ($\rho = 0$), though that is not a complete prescription of the conservationist optimality criterion.

It is convenient to introduce a modified Hamiltonian function—called the "current value" Hamiltonian—as follows:

$$H' = e^{\rho t}H$$
$$= e^{\rho t}U + e^{\rho t}P_K F$$
$$= U' + P_K' F \qquad (2.20)$$

where U' is no longer an explicit function of time, but P_K' now has an explicit time dependence. The new Hamiltonian equations are

$$\frac{\partial H'}{\partial \chi_t} = 0 \qquad (2.21)$$

$$\frac{\partial H'}{\partial K_t} = -\dot{P}_K' \qquad (2.22)$$

and

$$\frac{\partial H'}{\partial P_K'} = \dot{K}_t \qquad (2.23)$$

Notice that

$$\dot{P}_K' = \rho e^{\rho t}P_K + e^{\rho t}\dot{P}_K \qquad (2.24)$$

On carrying out the indicated operations, it can be verified easily that (2.21) and (2.23) reduce exactly to (2.12) and (2.1) as before, but (2.13) becomes

$$0 = \dot{P}_K + \frac{\partial U}{\partial K_t} + P_K \frac{\partial F}{\partial K_t} + \rho P_K \qquad (2.25)$$

which is identical to (2.15) except for the last term. Equation (2.25), which applies in the case where an exponential discount function is used, will generally be used in place of (2.15) in the following sections.

Formulating an actual model, of course, means selecting a control parameter χ_t, a social discount rate (or intergenerational transfer function), and a specific form for $F(K_t, \chi_t, t)$. In capital theory, the control parameter χ_t is often *consumption C*, reflecting the fact that capital investment is limited to the difference between total output and consumption. Total output is a given function (the "production function") of existing capital.[†] The utility or service function U, too, is some function of consumption. Thus the optimal path for capital accumulation can be characterized in terms of relationships between the production function and the utility function.

Extending capital theory to include natural (i.e., non-man-made) capital is conceptually quite straightforward. The roles of consumption and investment are basically unchanged. In the next two sections (2.2 and 2.3), the basic model described above is extended to nonrenewable and renewable resources, respectively. My discussion draws heavily

---

[†]The production function also involves other inputs, such as labor, which can be ignored for the moment.

from a useful review article by Fisher and Peterson.[11]

It is not the primary purpose of this chapter to criticize the body of theory that has grown from the use of these models, nor does the chapter have any pretensions to comprehensive coverage or detailed account of the derivations. Rather, the purpose is to provide an overview of barely sufficient breadth and depth to introduce the argument of Chapter 3 and the subsequent discussions of large-scale simulation and optimization models.

2.2   MODELS OF EXHAUSTIBLE RESOURCES

By general agreement, the "modern" (i.e., post-Malthusian) economic theory of exhaustible resources begins with the work of Harold Hotelling.[12] Hotelling constructed a simple model, using the calculus of variations, in which a nonunique, homogeneous, nonrenewable resource is owned by a private owner and asks two questions: How can the owner of the resource maximize his (private) return on that asset? and, What is the socially optimal rate of extraction? Hotelling considered both the case of a competitive free market for the output of the mine, characterized by a well-defined and well-behaved demand curve that is also known to the owner of the mine; and a monopolist mine owner who is able to influence demand through his control over prices.

Observing the analogy between man-made capital and natural capital, we can formulate a model for an optimal extraction strategy using the apparatus developed in the previous section. In brief, let K represent the "stock" of an exhaustible mineral resource and $\dot{K}$ its (negative) rate of extraction, which is assumed to be a function F of K itself, and a control variable χ, which can be interpreted as the intensity of mining "effort." Thus

$$K_t = K_0 + \int_0^t \dot{K}_\tau(K, \chi, \tau)\, d\tau = K_0 + \int_0^t F_\tau(K, \chi)\, d\tau$$

$$(2.26)$$

where the integral (which is negative) reflects cumulative extraction up to the time t. The current "yield" from the capital stock K remaining at time t is

$$U_t(K_t, \chi_t, t) = U_t(K_t, \chi_t)e^{-\rho t} \qquad (2.27)$$

where the time-preference function is explicitly

broken out. It is desired to maximize the present value of all future benefits from extraction of the resource, namely, to determine the conditions under which a "welfare" function W is given by which

$$W = \int_t^T U_\tau(K_\tau, \chi_\tau)e^{-\rho\tau}\, d\tau \qquad (2.28)$$

is maximized.

To represent the case of the mine-owner in a competitive market, it is convenient to express U_τ and F_τ as follows, omitting any explicit time dependence:

$$U_\tau(K_\tau, \chi_\tau) = P_\pi \Pi(K_\tau, \chi_\tau) - P_\chi \chi_\tau \qquad (2.29)$$

$$F_\tau(K_\tau, \chi_\tau) = -\Pi(K_\tau, \chi_\tau) \qquad (2.30)$$

where $\Pi(K, \chi)$ is the production function for the resource extraction activity (e.g., a mine), and χ_τ—which was previously identified as a variable controlling the intensity level of that activity—can now be thought of as "labor." Thus P_π is the shadow-price per unit of output of the mine and P_χ is the shadow price (i.e., the wage rate) of a unit of labor. Clearly, P_K is the shadow price—or marginal user cost—of a unit of ore still in the ground. Equation (2.29) identifies the mine-owner's utility function with the net profits of operating the mine, allowing for extraction costs.

The current value Hamiltonian for the problem now becomes

$$H' = P_\pi \Pi(K, \chi) - P_\chi \chi - P_K \Pi(K, \chi) \qquad (2.31)$$

and the conditions for maximizing W are given by applying (2.21) and (2.22), namely,

$$P_\chi = (P_\pi - P_K)\frac{\partial \Pi}{\partial \chi}(K, \chi) \qquad (2.32)$$

and

$$\dot{P}_K = \rho P_K - (P_\pi - P_K)\frac{\partial \Pi}{\partial K}(K, \chi) \qquad (2.33)$$

Substituting (2.32) in (2.33) and dividing by P_K one obtains

$$\frac{\dot{P}_K}{P_K} = \rho - \frac{P_\chi}{P_K}\frac{\partial \Pi/\partial K}{\partial \Pi/\partial \chi} \qquad (2.34)$$

where $\partial \Pi/\partial \chi$ and $\partial \Pi/\partial K$ are the marginal productivities of labor and capital stock, respectively.

The first condition can be rewritten to state that the price of the extracted ore P_π must be equal to the marginal labor cost of production—which is the price of a unit of labor P_χ divided by the output per unit of labor input, $\partial \Pi / \partial \chi$—plus the price of a unit of capital, which is to say, ore left in the ground. The second condition states that the price of unextracted ore P_K must increase over time, at a rate equal to the social discount rate ρ minus the marginal labor cost of production divided by the marginal capital cost of production. If the productivity of the mine is *not* a function of the amount of ore remaining, then $\partial \Pi / \partial K = 0$, and the price of ore remaining in the ground will optimally rise at the rate ρ, the market rate of interest. This is essentially the first result of Hotelling.[12] Incidentally, Hotelling went on to show that if U_τ is defined in such a way as to represent the consumer surplus associated with the output of the mine—as a reasonable surrogate for social utility—the optimal extraction rate is still such that the price will rise at the rate of market interest ρ.

But Equation (2.34) yields further results. If $\partial \Pi / \partial K$ were positive—which is plausible for a fishery or a forest under conditions of maximum sustained yield, but not for a mine—the optimal rate of increase of the value of the resource would be less than the social discount rate ρ, because the resource owner would benefit not only from present output but also from increased future production levels.[13] Conversely, when $\partial \Pi / \partial K < 0$, the productivity of the resource decreases with progressive exhaustion (which is more plausible for a mineral resource) and the optimal price P_K of resources left in the ground should rise faster than ρ.

In this model, it is implicitly assumed that prices are determined in a free market characterized by a known demand function. Actually, markets are often distorted in various ways, by government regulations or by oligopolistic industries. In particular, it is interesting to examine the case where the resource is controlled by a single seller—a simple monopolist—who can set the price to maximize his own benefit. Knowing the demand function, he sells to each consumer at exactly the "indifference" price, thus capturing part of the consumer surplus. Actually, it can be shown that when a mine-owner is in competition with other producers, "net price" to him means market price less marginal extraction cost. On the other hand, if he is a monopolist, he is interested in *marginal* revenue less *marginal* extraction cost. As is well known, the monopolist normally obtains a higher average net rate of return than a producer in a competitive market. But, on the other hand (some

unrealistic counterexamples notwithstanding[†]), the monopolist tends to hold his rate of extraction at a lower level than a competitive set of producers would do, thus conserving the resources for future use. Thus while the interest of short-term consumers lies in preserving competition to keep the market price as low as possible, the interest of future generations appears to be more consistent with that of monopolistic producers.

More generally, the Hotelling principle implies that if prices rise "too" slowly, resources will be extracted "too" fast (in terms of any given discount rate, or relative weighting of current and future interests), and conversely. It also implies that the market price will rise eventually along the demand curve to the point that demand is finally choked off. This should occur, assuming perfect markets in equilibrium, exactly at the moment the resource is physically exhausted—the perfect denouement!

It hardly needs to be pointed out that the Hotelling model simplifies the real world in numerous ways. First and foremost perhaps, one notes that markets are imperfect and real-world resource price behavior often reflects uncertainties and institutional rigidities that are fairly obvious to any observer. A rather significant departure from reality is that real resources are neither homogeneous in quality nor unique in source. (Moreover, many resources are by-products or co-products of others, which greatly complicates the supply–demand relationship.) A further key point is that empirical research by Barnett and Morse[14] (among others) suggests that outputs of most resource commodities seem to have been increasing for many decades, with declining market prices. This is in apparent contradiction to the Hotelling result. Further, Hotelling ignored the role of capital and the effect capital-related constraints will have on development strategy. Competition, entry, and departure of firms were not considered. Nor were the inherent uncertainties of exploration taken into account.

A related objection is that demand for the resource may be very inelastic in the short-run[‡] but

---

[†]Citations given in Reference 11.

[‡]For essentially technological reasons, an entire industry may be organized around a particular technology (e.g., the internal combustion engine), which, in turn, depends on certain inputs such as gasoline using tetraethyl lead (TEL). While new cars could be, and were, designed to use lead-free gasoline, TEL will be needed until all pre-1974 vehicles are scrapped—which might take 10 more years!

relatively elastic in the longer run. Thus the notion of a well-defined demand curve, constant over time, is unduly simplistic. In any event, the use of such a curve begs a fundamental question: What does *society* do when the resource is finally exhausted? An underlying assumption implicit in the use of such curves is that substitution is *always* possible on a time scale short enough such that society's actual survival could never be threatened by the absence of some resource. This puts a very high implicit reliance on the capabilities of technology—a reliance that many scientists and technologists themselves are beginning to question.

One last simplification of importance concerns the noncongruence of private and public utilities and, by extension, the difference between private and public discount rates. In part, this lack of congruence arises from the existence of significant and pervasive market failures, or externalities.[†] One such source of externalities is the extraction process itself and the associated wastes and nuisance; subsequent processing of the resource commodities also yields harmful residuals, as does "final" consumption and ultimate discard. In another sense, it arises from the fact that environmental benefits are largely indivisible and consequently not exchangeable in a marketplace, even in principle. The same comment applies, incidentally, to the externalities associated with resource exhaustion: The losses clearly extend beyond the nexus of immediate producers and direct consumers. One key reason (among several) is that *potential consumers in future generations cannot bid against current consumers in any marketplace.*[‡] The question of optimal resource extraction in the presence of environmental constraints on the assimilation of residuals is deferred to the next section.

An obvious extension of Hotelling's model is to discard the assumption of a single homogeneous source and introduce the possibility of multiple sources of different qualities.[§] The question im-

mediately arises: Which should be exploited first? Extending Hotelling's model to incorporate a second source of different quality (i.e., extraction cost), it was shown by Herfindahl that only the cheaper can produce at one time.[17] In due course, the net price (and the market price) will rise and output will decline. At a certain moment in time the first source is exhausted: at precisely this moment (and price) the second producer should enter the market, and so on. The Herfindahl model has been extended to include a continuous spectrum of sources, free entry, and exit of mining enterprises, the possibility of recycling, technological change, and various other refinements, by Schultze.[18] Schultze does not consider recycling as a form of waste treatment for environmental management purposes, but rather as a means of extending the life of exhaustible resources by "mining" stocks of discarded wastes. The latter are regarded simply as a type of ore. However, the complexities of co-product relationships and time-dependent elasticities of substitution have not yet been given adequate attention in the context of the Hotelling model.

Regarding the problem of erratic commodity price behavior, there are both simple and complicated explanations. A simple one is that the market is highly imperfect and that most of the transactors are forced to operate without full information. As a consequence, there are opportunities for manipulation and "boom–bust" cycles of speculative hoarding in rising markets followed by forced dumping in falling ones. The question is whether these market imperfections are so severe as to invalidate the use of market prices as an allocative mechanism. Solow has argued that destabilizing tendencies in the commodity markets may be compensated, in part, by corrections in the asset market.[19] In other words, if commodity prices are "too low," asset owners might be tempted to convert their assets into cash by increasing production, thus exacerbating the situation. However, because of the existence of a market for assets, the market value of the asset will simply drop, automatically raising the rate of return. This mechanism is real enough, but it has not been incorporated in a model.

The role of capital as a determinant of extractive industry (or firm) strategy was stressed by V. L. Smith,[20] although his treatment was imperfect. An alternative model for a single firm, including capital, was presented by Cummings and Burt,[21] the principal result being that the optimal strategy is one in

---

[†]The literature of externalities is extensive, but until recently, as Mishan has pointed out, most of the contributions treated externality effects as a comparatively minor deviation from Pareto optimality.[15] Ayres and Kneese seem to have been the first to recognize the pervasive nature of externalities associated with materials flow and residuals generation.[16]

[‡]This is another example of a "pervasive" externality.

[§]Hotelling himself considered the possibility of a single source such that marginal cost of extraction increases with time. This is equivalent to a situation in which the quality of the resource varies. In such a case, the only feasible strategy is to remove the highest quality ore first.

which the immediate marginal cost of capital invest-ment is equal to the discounted value of income generated by it in the future through its impact on a profit function.

The apparent discrepancy between real-world pro-ducer behavior (increasing output at declining prices) and the behavior called for by the Hotelling principle has not been completely laid to rest. A combination of modifications of the simple Hotell-ing model would explain it, however. To start with, one can suppose that demand for commodities (at a given price) is increasing faster than the market rate of interest, because of economic growth. In effect, the demand curve is not constant in time, but is moving to the right in such a way that it intersects the upward swinging supply curve at a point corre-sponding to a larger demand, (but also at a higher price). To get around this problem, a further modifi-cation is needed, namely, declining production costs (due to improved extraction technology).

Schultze has introduced the cost of extraction as an explicit variable in his model,[18] and he discusses the model modifications required to make this a decreasing function of time. He also discusses, briefly, the implications of an exponentially decreas-ing cost. The effect is to replace the Hotelling result —a scarcity rent for mineral rights that increases at the rate of market interest—by a U-shaped price function that declines before eventually rising. Moreover, major new discoveries (or technologies) may have a discontinuous impact, forcing the fu-tures market to "re-initialize" occasionally. This can result in successive exponential rising price curves starting from successively lower plateaus. However, Schultze notes that complex behavior such as this deserves more extended treatment than it has yet received, in a model that would also take account of exploration costs, discovery probability, reserves, risk, and so on. A schematic comparison of the models of Hotelling, Herindahl, Cummings and Burt, and Schultze is given in Figure 2.1.

In passing, it might be noted that a recent econo-metric study of the petroleum drilling industry, by Richard B. Norgaard, showed that the rate of tech-nological change, leading to decreasing unit extrac-tion costs, is very much greater than had previously been thought.[22] Norgaard explicitly considers a number of factors that would otherwise result in increasing costs, such as the declining success rate for newly drilled wells, the effect of increasing depth and exploiting more difficult formations. Without

compensating technological changes, these factors combined would have increased well-drilling costs by 233% between 1939 and 1968 in the United States. Yet wells increased only 64% in real cost during this time period. Assuming deposits of equal quality, well-drilling costs decreased between 53% and 87% over the period.

Another deficiency of the simple Hotelling model is that it assumed (implicitly) that mineral resources are essentially equivalent to *inventories*; that is, that their location and quantities are fully known. In reality, the situation is quite different. The U.S. Bureau of Mines classifies resources as *reserves;* and *conditional, hypothetical and speculative resources*—as shown in Figure 2.2—according to the degree of certainty of existence and cost of recovery.[23] There is a great deal of doubt and dispute over the magni-tudes belonging in each category, especially hypothetical and speculative (i.e., undiscovered) re-sources. A major and increasing cost intimately associated with resource extraction is the discovery and "mapping" of underground deposits.

If exploration to eliminate uncertainty is indeed not an important element of cost to mining firms, then their response to a rise in market interest rates —a surrogate for the social discount rate—"should" stimulate increased output, and conversely.[24-26] Adelman[27] and Peterson,[28] among others, have ex-plored the relationship between discount rates and exploration per se, concluding generally that short-term and long-term behavior may differ, with higher discounts leading to initial over-exploitation followed by over-conservation. Common property attributes of information resulting from exploration —resulting in unearned windfall gains by land-owners, for instance—contribute to this phenome-non.

Uncertainty about future technological substitu-tions is yet another pertinent aspect of the subject. The case of a known substitute becoming available at an unknown future time has been considered by Dasgupta and Heal.[29] The social discount rate ap-pears to be an inadequate means of reflecting uncer-tainties of this kind. Dasgupta and Stiglitz have explored the same scenario more thoroughly under a variety of alternate assumptions regarding owner-ship and competition.[30]

The difficulties caused by lack of congruence be-tween private and public utility functions and dis-count rates have recently received attention from a number of theorists (including Page, mentioned

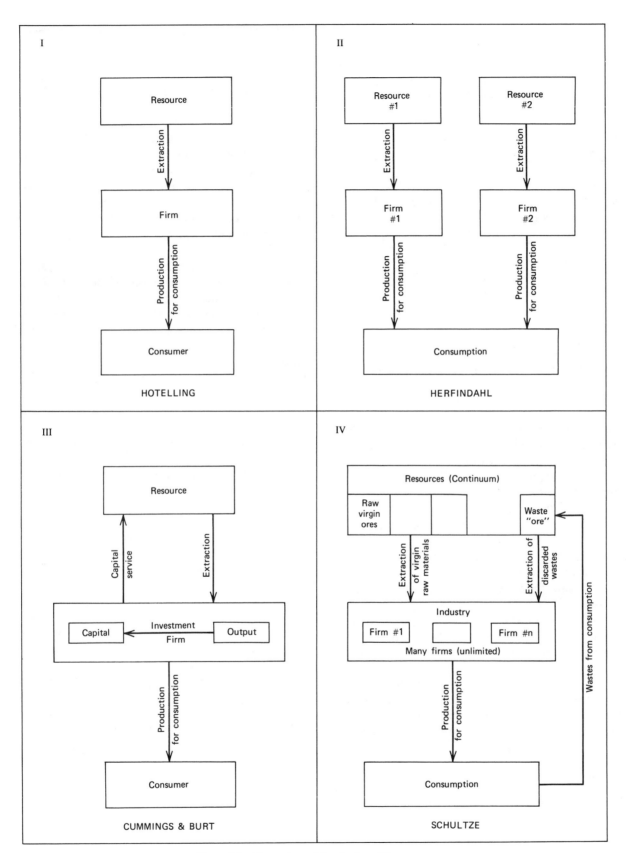

Figure 2.1. Dynamic models of exhaustible resources.

27

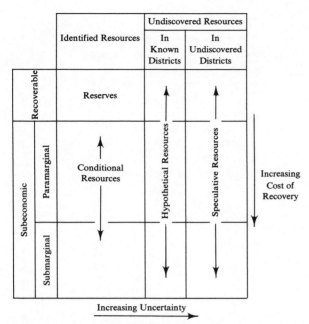

Figure 2.2. Resource classification.

earlier) with some interesting results. Hotelling argued, in effect, that the market rate of interest is a reasonable surrogate for a social discount rate. However, this assumption seems increasingly questionable, for a number of reasons.

A few examples will suffice. The most widely used formulation of utility in economics, by Ramsey,[31] introduced the sum of utilities for periods, discounted such that each successive period is given a lower weight than the previous one. The rate of discount selected may vary from zero up to the market rate of interest, depending on who makes the choice. Also, the time horizon over which the sum-over-periods is extended can be finite or infinite. There has been considerable controversy over these points since (as it turns out) the choice of utility function essentially determines the results of the analysis.

It has been shown by Solow[32] and by Dasgupta and Heal[29] that with any positive discount rate the present is favored over the future to such an extent that per capita consumption declines asymptotically to zero in the optimal (utility-maximizing) path.[†] On

---

[†]This conclusion is dependent on the assumption about substitutability between renewable and exhaustible resources. If there is an upper limit on technological improvement, disaster eventually follows exhaustion. And conversely, if the elasticity of technological substitution exceeds unity and the elasticity of output with respect to reproducible capital exceeds the elasticity of output with respect to exhaustible resources, then a perpetual steady state is possible with nonzero consumption, depending on population and initial capital.

the other hand, with a zero discount rate, and a finite resource base, consumption per head would be constant until the resource is exhausted. If "existence" per se has high utility, the consumption level will be held at the minimum level consistent with survival. This is a significant result because, in the absence of a resource problem, the optimal path rises to a consumption plateau ("bliss") much higher than current levels, even if the future is quite heavily discounted. In that case, favoring the present over the future seems entirely reasonable since our descendents would, in any case, be much richer than we are.

A final special example, discussed by d'Arge,[33] is noted because it leads naturally to the next section, where we consider the effect of incorporating environmental elements, including regeneration and pollution, in the resource model. The example is of an astronaut—or it could just as well be a group of astronauts—in a sealed spaceship "lost in space" with no hope of rescue. Question: How should he (they) optimally allocate the limited supply of provisions? Assume, first, that the availability of food (i.e., resources) is the only limiting factor and that the supply is, in any case, insufficient for a normal lifespan. In this situation, the result depends how the astronaut weights a unit of current consumption vis-à-vis a unit of life extension. If the latter is paramount, he will reduce consumption to or near the subsistence minimum and survive as long as he can. Even if extended life has less than infinite value, a diminishing marginal utility of consumption would prevent him from consuming much above the minimum level.

But, now suppose the accumulating metabolic wastes cannot be discharged or recycled, and suppose that this adversely affects the astronaut's utility level. In that case, if life extension does not have infinite value, sooner or later the astronaut will sacrifice some time at the end of his life in order to increase current consumption.[†] His optimal path will be to consume as little as possible at first (to minimize the rate of waste accumulation) and to consume more rapidly later, to compensate for the increasing discomfort of his polluted environment and maintain a constant level of "real" enjoyment.

The interesting point about the last parable is that it leads to a counter-Hotellian result. The circumstances are certainly different—which accounts for

---

[†]Cigarette smokers, heavy drinkers, and heavy eaters often consciously make a similar trade—though without the element of absolute certainty that the astronaut would have.

the different optimal path—but a certain similarity between the spaceship and the mine remains. A utility function that gives significant weight to the future, in combination with an increasing disamenity associated with cumulative consumption (= extraction) can lead to a completely different optimal policy than the one deduced from a Hotelling model.

2.3 DYNAMIC MODELS OF RENEWABLE RESOURCES

Although dynamic models of renewable resources —such as fisheries—have been appearing regularly since (at least) the work of the early mathematical biologists Raymond Pearl[34] and Alfred Lotka,[35] it is only quite recently that the problem was clearly perceived to be an application of capital theory.[10] As a matter of fact, one can regard a fishery, a forest, or an agricultural system very much like a mine except that the former is capable of natural regeneration. For an exhaustible resource, it will be recalled [Equations (2.30) and (2.26)], we defined

$$\dot{K}_t = F(K, \chi, t) = -\Pi(K, \chi)$$

and

$$K_t = K_0 + \int_0^t F(K, \chi, t)\, dt$$

since the rate of depletion of the capital stock can be identified with the physical output of the mine. In the case of a biological system, however, an additional term must be added to reflect the regenerative function. Thus Equation (2.30) can be generalized to the following form

$$F_\tau(K_\tau, \chi_\tau) = g(K_\tau) - \Pi(K_\tau, \chi_\tau) \qquad (2.35)$$

where $g(K_\tau)$ is a natural "growth" term. More precisely, $g(K_\tau)$ reflects *net* natural capital increase, which is a function of the size of the stock of natural capital K. In all biological systems, there is an ultimate maximum value for K (e.g., population) where the net rate of increase vanishes, due to overcrowding, increased competition for limited food supplies, increased predation, or some combination of these factors. Thus the closer K approaches to its maximum value, the slower its rate of increase. On the other hand, when K is very small, its absolute rate of increase is also likely to be small in (rough) proportion to K itself. Thus one would expect $g(K)$ to

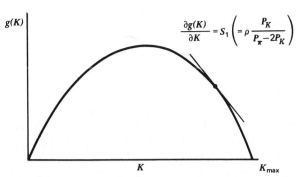

$$\frac{\partial g(K)}{\partial K} = S_1\left(= \rho\, \frac{P_K}{P_\pi - 2P_K}\right)$$

Figure 2.3. Growth as a function of stock.

vanish for $K = 0$ and for $K = K_{max}$, reaching a maximum value for some intermediate value. The situation is depicted graphically in Figure 2.3.

In the absence of any harvesting by Man, the growth pattern of the capital stock (e.g., of fish or trees) would be determined by the differential equation

$$\dot{K} = g(K) \qquad (2.36)$$

which is directly integrable

$$\int_0^K \frac{dK}{g(K)} = \int_0^t d\tau = t \qquad (2.37)$$

The integral can be expressed in closed form for many simple functions $g(K)$. Qualitatively, if $g(K)$ has the form shown in Figure 2.3, it is easy to see that K, a function of time (absent harvesting), will take the form of an elongated S-curve, as shown in Figure 2.4.

The simplest form of $g(K)$ having the desired properties was introduced by Pearl[34] to explain the growth of a population in a limited environment

$$g(K) = \alpha K(K_{max} - K) \qquad (2.38)$$

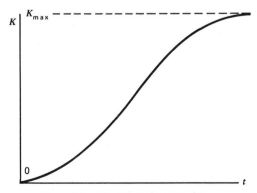

Figure 2.4. Capital stock versus time.

It leads to the familiar "logistic" form for $K(t)$

$$K(t) = K_{\max}\big[1 + \exp(-\alpha t)\big]^{-1} \qquad (2.39)$$

Obviously, without regeneration, any finite stock of resources will be depleted by continuous extraction, no matter how low the rate. However, when there is regeneration, the outcome is uncertain: The resource *may* be exhausted eventually, depending on the harvesting strategy. Thus the renewable resource model touches on some very interesting (and important) issues, including the optimal rate of extraction vis-à-vis the maximum sustainable yield, and the conditions under which "exhaustion" (i.e., extinction) may occur.

The Hamiltonian for a renewable or regenerating resource is

$$H = U + P_K F$$

$$= \big[P_\pi \Pi - P_\chi \chi\big]e^{-\rho t} + P_K(g - \Pi) \qquad (2.40)$$

and [using (2.20) et seq.] the conditions for an optimum path yield Equation (2.32) as before

$$P_\chi = (P_\pi - P_K)\frac{\partial \Pi}{\partial \chi} \qquad (2.32)$$

and

$$\dot{P}_K = \rho P_K - (P_\pi - P_K)\frac{\partial \Pi}{\partial K} + P_K\frac{\partial g}{\partial K} \qquad (2.41)$$

Using (2.32) again, we get

$$\frac{\dot{P}_K}{P_K} = \rho - \frac{P_\chi}{P_K}\frac{\partial \Pi/\partial K}{\partial \Pi/\partial \chi} + \frac{\partial g}{\partial K} \qquad (2.42)$$

which is the same as (2.34) except for the last term.

The Equations (2.32) and (2.42) can be manipulated in various ways to yield some interesting results. A helpful way of visualizing the behavior of the optimal path under alternative assumptions is to consider the steady-state conditions.† From the condition for an unchanging (steady-state) capital stock,

$$\dot{K} = F(K) = 0$$

---

†There is quite extensive literature on "turnpike" theorems, which define the conditions under which it is safe to say that an optimal path is "usually" close to a steady-state solution. At any rate, this is a convenient assumption.

whence [from (2.35)] the growth rate must exactly equal the yield

$$g(K_t) = \Pi(K_t, \chi_t) \qquad (2.43)$$

from an optimal capital stock K.

Differentiating and equating both sides of (2.43) with respect to K and setting $\dot{P}_K = 0$ in (2.41), we obtain a steady-state relationship, namely,

$$\frac{\partial \Pi(K, \chi)}{\partial K} = \frac{\partial g}{\partial K} = \rho\frac{P_K}{P_\pi - 2P_K} \qquad (2.44)$$

Equation (2.44) states that the optimal steady-state output of a renewable source does *not* normally occur at the point of maximum sustainable yield, where the growth function $g(K)$ has its peak value (and zero slope), namely,

$$\frac{\partial g}{\partial K} = 0$$

Optimum steady-state exploitation only coincides with maximum sustained yield in the special case of a zero social discount rate ($\rho = 0$). For positive values of ρ, which imply that a unit of future production is valued somewhat below a unit of present production, the optimum yield occurs for a positive slope of $g(K)$, as suggested in Figure 2.3, where S is the optimum slope.

The other condition for an optimum is obtained from (2.32)

$$\frac{\partial \Pi(K, \chi)}{\partial \chi} = \frac{P_\chi}{P_\pi - P_K} \qquad (2.45)$$

If one plots yield $\Pi(K, \chi)$ as a function of effort (labor), it is evident that there is a family of curves for increasing values of stock K, each one of which has the optimum slope (2.45) at *some* K, as shown in Figure 2.5. The locus of all such intersections defines the optimal level of extraction effort as a function of K. The optimal value for K itself can be determined by a similar graphical exercise using Figure 2.3. Note that, as a rule, the optimal slope for $\partial g/\partial K$ would be negative, although under certain conditions (very high prices for current output in relation to the value of future production) it might be positive. A positive slope for $\partial g(K)/\partial K$ would imply a stock too small for maximum regeneration in the absence of extraction. In extreme cases, it is possible that harvesting *can* be so intensive that a resource

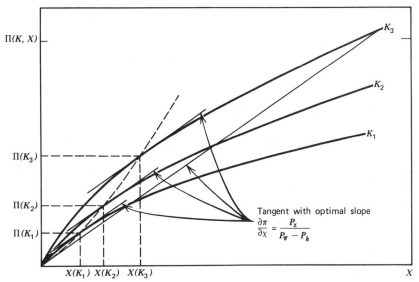

Figure 2.5. Optional production as a function of stock and labor.

such as a species of fish becomes unviable (i.e., natural reproduction fails to compensate for natural losses). The ownership of the resource is quite an important factor. It is possible that economic consideration may dictate the exploitation of a renewable resource to the point of extinction, even if the resource has a single owner. This outcome is much more likely, however, if there is no owner at all—the resources are common property. These issues have been examined in depth by numerous authors.[36-46]

Although the foregoing analytic framework is fairly general, a glance at the referenced articles suggests that most of the relevant research was done with fisheries in mind. The problem of ownership is crucial in this instance. Fish are the classic example of a "common property resource" whose benefits accrue to any and all who can appropriate them. Another important example of common property resource, of course, is the waste assimilative capacity of the environment. I now shift attention to the latter, inasmuch as some new points arise.

2.4 ABSTRACT MODELS OF ENVIRONMENTAL EXTERNALITIES

The literature of economic models of environmental externalities seems to have a much shorter history than the literature of exhaustible resources. Some of the basic ideas of environmental economics—particularly the problems arising from the fact that Man's activities are confined to a fixed and finite environment from which he derives a number of essential services, including that of waste assimilation—were first formulated in economic language by Boulding.[47] An early attempt to formulate a (static) general equilibrium model seems to have been that of the author and A. V. Kneese.[16, 48] The key point in this paper was that environmental externalities are inherent and pervasive consequences of the extraction, processing, "consumption," and discharge of physical materials and energy. Apart from (temporary) accumulation in the form of capital goods, the laws of conservation of matter and of energy guarantee that the sum total of all waste flows to the environment from the economy must be equal to the sum total of all resources originally extracted from the environment. This rule—known as the materials/energy balance principle—also applies to individual communities, regions, firms, or industries.

The paper also proposed an elaboration of the Walras–Cassel[49] general equilibrium model and showed that if a static equilibrium exists for the Walras model, it should also exist for the extended version. The A–K model resembles, in some respects, a closed input–output model with physical terms, in which the "environment" is treated as a distinct sector so that all physical flows balance. However, it involves two successive transformation matrices: a resource/commodity matrix and a commodity/product matrix. Products are divided into separate goods and services categories; the latter involve no physical inputs. Market prices are attached to commodities produced by the economic sectors and raw materials extracted from the environment. The return flow of waste residuals from final consumption back to the environment also carries a "price," but the price is negative—we can imagine an environmental agency charging a fee for disposal. Recycling is allowed for.

The major drawbacks of the A–K formulation are:

1. The model is static.
2. It assumes fixed production technologies and, consequently, fails to allow for the explicit possibility of technological substitution or optimal

choice of inputs. The problem of generalizing the model to allow for technological choice—as applied to environmental or other optimization criteria—is taken up later in the book. The remainder of this chapter considers other recent developments related to modeling environmental externalities.

A deeper and more extensive examination of these issues can be found in a recent monograph by Karl-Göran Mäler.[50] Mäler extends the A–K general equilibrium formulation into a somewhat more complex static model (see Figure 2.6). He also provides

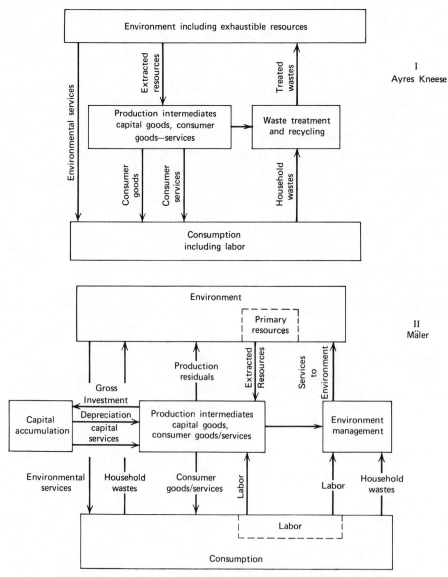

Figure 2.6. General equilibrium models, including environmental links.

detailed rigorous proofs of the existence of a competitive equilibrium, assuming "charges" are imposed for environmental services, including the acceptance of wastes. It should be noted in passing that Mäler's general equilibrium model does not strictly conform to the materials balance requirement, since physical inputs and wastes accepted by the environmental management sector are mapped into an undifferentiated service.

While the A–K and Mäler formulations provide a formal link, through the materials/energy balance principle, between resource extraction and environmental externalities, they do not take the further step of seeking optimal policies for economic growth in the presence of resource constraints and a degradable environment—where the degradation is a function of the rate of extraction and consumption. Two recent contributions have taken this additional step, thereby providing a formal bridge between the older literature of optimal growth and optimal extraction models and the newer interest in environmental externalities.

Probably the first integrated treatment of the environment as a kind of exhaustible resource has been provided by d'Arge and Kogiku.[51] They first construct a simple optimal path model of the Ramsey type with a single material consumer good. Waste is assumed to be a joint product of the output of this good (utilizing the materials balance principle and neglecting the possibilities for recycling). In a fixed environment, it follows that growth in material consumption leads a corresponding increase in waste flows. If there is no environmental assimilation (or if the rate of assimilation is less than the flow of wastes) the density of cumulative pollution will inevitably rise. This is a transparent Malthusian conclusion of no special interest.[†] However, d'Arge and Kogiku then introduce a utility function, depending on both the level of consumption and the density of wastes, consisting of a sum of utilities for successive time periods—i.e., generations—with varying discount rates. A finite time horizon is assumed, with the length of the interval regarded as part of the objective function. Using the theory of optimal controls they then investigate the optimum (utility-maximizing) path of economic growth over time as a function of the assumed rate of discount of future

time periods as compared to the present. The major result is that consumption should be sufficient to equate the marginal utility (or shadow price) of *consumption* to the marginal disutility of the associated *waste output*, which is assumed to increase as the density of accumulated wastes increases. Consequently, optimal consumption is initially lower when the pollution level is lower; later it rises as people consume faster to compensate for the disamenity of a high cumulative pollution level.

The d'Arge–Kogiku model was also extended to consider the more complex case that arises when both renewable and nonrenewable resources are taken into consideration, and waste output is made a function of the rate of extraction. In this case, the optional rate of *extraction* (to which consumption is related, of course) is such that the marginal utility of resource extraction—in terms of consumption permitted thereby—is equal to the marginal disutility of the associated waste, as before. The possibility of a recycling has also been explored by d'Arge and Kogiku using a variant of the model.

Mäler's monograph[50] includes an extensive analysis of optimal consumption in a similar model. Again, there is a finite stock of resources whose extraction results in consumption of capital accumulation, both leading eventually to production of wastes. The environment is assumed to generate a service, which is degraded by the accumulation of wastes. Unlike the d'Arge–Kogiku case, some assimilative (i.e., recycling) capacity is assumed, so that if the rate of discharge of residuals is held at a low enough level, the ambient level of wastes will not increase.

Utility in the Mäler model is assumed to depend on the level of consumption and the level of environmental services—which is, of course, a function of the level of wastes. The overall utility function is the sum of discounted single period utilities, over a finite horizon or "planning interval."

The model can be used to assess the differences between optimal paths in cases where the assimilative capacity is very small or nonexistent (d'Arge and Kogiku) vis-à-vis cases where it is large. In the former case, Mäler points out that the limitation of stock of exploitable natural resources may not be an actual constraint, since extraction and consumption will be discouraged by environmental disamenity costs. However, if assimilation capacity is large, extraction and consumption will be less inhibited and resource scarcity may be the limiting factor.

[†]Both the assumption that total consumption is equal to materials consumption and the assumption that waste is a joint product of material consumption are evidently unrealistic.

In addition to the simple example, Mäler considers extensions of the model in which there are two or more resources, increasing population and recycling. The Mäler and d'Arge–Kogiku models are compared schematically in Figure 2.7.

Several other authors[52–54] have also treated the problem of optimal waste accumulation as a direct analog of capital theory. In this approach, it is postulated, in effect, that the utility function depends upon the accumulated *stock* of pollution (rather than the flow), and the rate of change of the pollution stock is a function of current consumption resources devoted to abatement and the current flow. A natural rate of waste decay can also be assumed in analogy with the natural growth rate of a renewable resource. As before, the conditions for an optimum accumulation can be expressed as differential equations, and socially efficient "rules" for waste management can be derived. The general conclusion is that, on an optimal path, the marginal utility of consumption is equal to the marginal cost of production plus the marginal disutility of the accumulated wastes.

In all optimal path models, the high degree of aggregation of the variables virtually nullifies any possible reason for obtaining actual numerical solutions. As noted earlier, the simple models of the type considered in this chapter are essentially used only to generate theorems. Unfortunately, if the models are made sufficiently complex to begin to be capable of simulating "real-world" behavior, they rapidly become mathematically intractable.

REFERENCES

1. F. P. Ramsey, "A Mathematical Theory of Saving," *Eco. J.*, **38**, 543–559 (December 1924).

2. H. Hotelling, "A General Mathematical Theory Of Depreciation," *J Amer. Stat. Assoc.*, **20**, 340–353 (September 1925).

3. K. E. Boulding, "The Economics of the Coming Spaceship Earth," in H. Jarrett (ed.), *Environmental Quality in a Growing Economy*, Essays from the Sixth Resources for the Future Forum, Johns Hopkins University Press, Baltimore, 1966.

4. R. Bellman, *Dynamic Programming*, Princeton University Press, Princeton, NJ, 1957.

5. L. S. Pontrjagin, V. G. Boltyanskii, R. V. Gamrelidze, and E. F. Mishchenko, *The Mathematical Theory of Optimal Processes*, K. N. Trirogoff (trans.), New York, 1962.

6. K. J. Arrow, "Discounting and Public Investment Criteria," in A. V. Kneese and S. C. Smith (eds.), *Water Research*, Washington, DC.

7. K. J. Arrow, "Applications of Control Theory To Economic Growth," *Mathematics of the Decision Sciences*, Part 2, American Mathematical Society, Providence, RI, 1968.

8. K. J. Arrow and M. Kurz, "Public Investment, the Rate of Return and Optimal Fiscal Policy," Stanford University Institute for Mathematical Studies in the Social Sciences, Stanford, CA, 1968.

9. R. Dorfman, "An Economic Interpretation of Optimal Control Theory," *Amer. Econ. Rev.*, 817–831 (December 1969).

10. T. Page, *Conservation and Economic Efficiency—An Approach to Materials Policy*, Resources for the Future, Johns Hopkins University Press, Baltimore and London, 1977.

11. F. M. Peterson and A. C. Fisher, "The Economics of Natural Resources," University of Maryland, mimeo, 1976.

12. H. Hotelling, "The Economics of Exhaustible Resources," *J. Polit. Econ.*, **39**, 137–175 (April 1931).

13. R. G. Cummings, "Some Extensions of the Economic Theory of Exhaustable Resources," *West. Econ. J.* **7** (September 1969).

14. H. Barnett and C. Morse, *Scarcity and Growth*, Johns Hopkins Press, Baltimore, 1965.

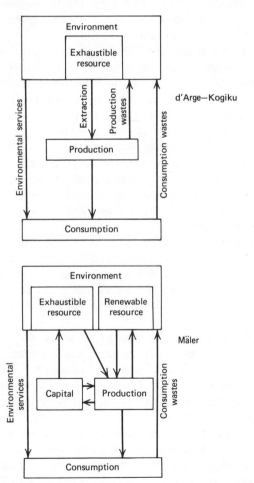

Figure 2.7. Optimal consumption models, including environmental links.

15. E. J. Mishan, "Reflections on Recent Developments in the Concept of External Effects," *Canad. J. Econ. Polit. Sci.* **31** (February 1965).

16. R. U. Ayres and A. V. Kneese, "Production, Consumption and Externalities," *Amer. Econ. Rev.*, **59** (3) (June 1969).

17. O. C. Herfindahl, "Depletion and Economic Theory," in M. Gaffney (ed.), *Extractive Resources and Taxation*, University of Wisconsin Press, 1967.

18. W. Schultze, "The Optimal Use of Non-Renewable Resources: The Theory of Extraction," *J. Environ. Econ. Management*, **1** (1) (May 1974).

19. R. M. Solow, Richard T. Ely Lecture: "The Economics of Resources or the Resources of Economics," *Amer. Econ. Rev.*, **64** (2) (May 1974).

20. V. L. Smith, "Economics of Production from Natural Resources," *Amer. Econ. Rev.*, **58**, 409–431 (June 1968).

21. R. G. Cummings and O. R. Burt, "The Economics of Production from Natural Resources: Note," *Amer. Econ. Rev.*, **59**, 985–990 (December 1969).

22. R. B. Norgaard, "Resource Scarcity and New Technology in U.S. Petroleum Development," *Nat. Resources J.*, **15** (2) (April 1975).

23. U.S. Bureau of Mines, *Minerals Facts and Problems*, U.S. Government Printing Office, Washington, DC, 1970.

24. J. Ise, "The Theory of Value as Applied to Natural Resources," *Amer. Econ. Rev.*, **15** (June 1975).

25. T. C. Koopmans, "Some Observations in Optional Economic Growth and Exhaustible Resources," in H. C. Bos, H. Linnamann, and P. de Wolff, *Economic Structure and Development*: *Essays in Honor of Jan Tinbergen*, North-Holland, Amsterdam, 1973.

26. A. C. Fisher and J. V. Krutilla, "Resource Conservation, Environmental Preservation and the Rate of Discount," *Quart. J. Econ.*, **89** (3) (August 1975).

27. M. Adelman, *The World Petroleum Market*, Johns Hopkins University Press, Baltimore, 1972.

28. F. Peterson, "A Theory of Mining and Exploring for Exhaustible Mineral Resources," University of Maryland, mimeo, 1975.

29. P. Dasgupta and G. Heal, "The Optimal Depletion of Exhaustible Resources" *Rev. Econ. Stud.*, *Symposium on the Economics of Exhaustible Resources*, 3–28 (1974).

30. P. Dasgupta and J. E. Stiglitz, "Uncertainty and the Rate of Extraction under Alternative Institutional Arrangement," Tech. Report No 179 Economics Series Institute for Math Studies in Social Sciences Stanford University, March 1976.

31. F. Ramsey, "A Mathematical Theory of Saving," *Econ. J.*, **38** (1928).

32. R. M. Solow, "Intergenerational Equity and Exhaustible Resources," *Rev. Econ. Stud.* (1974).

33. R. d'Arge, "Economic Growth and the Natural Environment," in A. V. Kneese and B. Bower (eds.), *Environmental Quality Analysis—Theory and Method in the Social Sciences*, Johns Hopkins University Press, Baltimore, 1972.

34. R. Pearl, *The Biology of Population Growth*, Alfred Knopf, New York, 1925.

35. A. J. Lotka, *Elements of Mathematical Biology*, 1925, reprint, Dover Editions, New York, 1956.

36. M. B. Schaefer, "Some Aspects of the Dynamics of Populations to the Management of the Commercial Marine Fisheries," *Bull. Interamer. Tropical Tuna Commission* (1954).

37. M. B. Schaefer, "Some Considerations of the Population Dynamics and Economics in Relation to Commercial Marine Fisheries," *J. Fish Res. Board of Canada* (September 1957).

38. C. Plourde, "A Simple Model of Replenishable Natural Resources Exploration," *Amer. Econ. Rev.*, **60** (*June* 1970).

39. V. L. Smith, "Economics of Production From Natural Resources," *Amer. Econ. Rev.*, **58** (3) (June 1968).

40. V. L. Smith, "On Models of Commercial Fishing," *J. Polit. Econ.*, **77** (2) (March/April 1969).

41. A. D. Scott, "The Fishery: The Objective of Sole Ownership," *J. Polit. Econ.*, **63** (April 1955).

42. J. Crutchfield and A. Zellner, "Economic Aspects of the Pacific Habitat Fishery," *Fish. Indus. Res.* **1** (1) (April 1962).

43. G. M. Brown, "An Optimal Program for Managing Common Property Resources with Congestion Externalities," *J. Polit. Econ.*, **82** (1) (January/February 1974).

44. J. P. Quirk and V. L. Smith, "Dynamic Models of Fishing," in A. D. Scott (ed.), *Econ. Fish. Management*; *A Symposium*, Vancouver University, British Colombia, 1970.

45. C. W. Clark, "Profit Maximization and the Extinction of Animal Species," *J. Polit. Econ.*, **81** (4) (July/August 1973).

46. P. Neber, "Notes on the Volterra Quadratic Fishery," *J. Econ. Theory*, **8** (1) May 1974.

47. K. Boulding, "The Economics of the Coming Spaceship Earth," in H. Jarrett (ed.), *Environmental Quality in a Growing Economy*, Johns Hopkins Press, Baltimore, 1966.

48. A. V. Kneese, R.U. Ayres, and R. C. d'Arge, *Economics and the Environment—A Materials Balance Approach*, Johns Hopkins University Press, Baltimore, 1970.

49. L. Walras, *Elements d'Economic Politique Pure* (1877), Jaffe Translation, London (1954). Also, G. Cassel, *The Theory of Social Economy*, New York, 1932.

50. K-G. Mäler, *Environmental Economics—A Theoretical Inquiry*, Johns Hopkins University Press, Baltimore, 1974.

51. R. C. d'Arge and K. C. Kogiku, "Economic Growth and the Environment," *Rev. Econ. Stud.*, **XL** (1) (January 1973).

52. E. Keller, M. Spence, and R. Zeckhauser, "The Optimal Control of Pollution," *J. Econ. Theory*, **4** (1) (1972).

53. C. G. Plourde, "A Model of Waste Accumulation and Disposal," *Canad. J. Econ.*, **5** (1972).

54. V. L. Smith, "Dynamics of Waste Accumulation: Disposal Versus Recycling," *Quart. J. Econ.*, **86** (4) (1972).

CHAPTER THREE Application of Physical Principles to Economics

The purpose of this chapter is twofold. First, it is intended to indicate some ways in which established physical principles intersect economic theory and, especially, note some important consequences that appear to have been unduly neglected in the past. Second, the chapter aims to provide a sort of bridge from the abstract concepts of "pure" economic theory, such as were discussed in the previous chapter, to the more detailed complex and empirically substantiated constructs that are needed for practical implementation of realistic large-scale simulation or optimization models. Lest a reader object that these are really two different subjects, it should be emphasized again that the principal purpose of this book is to introduce and describe a particular group of empirical models. The discussion of abstract theory is essentially orientational rather than definitive.

Regarding the applicability of physical principles to economics, three specific topics deserve discussion. First, in Section 3.1 we consider the implications of the first law of thermodynamics: the conservation of mass and energy. Some of these implications have already been discussed in the previous chapter, but others have not. Second, in Section 3.2 we consider the implications of the second law of thermodynamics, better known as the principle of "increasing entropy" or "increasing disorder." Although this principle has frequently been referred to, and occasionally invoked (sometimes incorrectly)

by economists, its specific implications for economic theory have never been dealt with systematically.[†]

The next section (3.3) concerns the structural implications for economics of specific *properties* of physical materials and forms of energy and the laws governing their transformations and their substitutions. These attributes and laws also determine the performance of structures, machines, chemical processes, or electronic/communications systems of various kinds. All physical capital (equipment and structures), whether it is used for extracting, manufacturing, or generating "final" services for consumers, is composed of systems, subsystems, components, and (ultimately) materials. This topic is both inchoate and immense, but the time seems ripe to

---

[†]This statement is made notwithstanding the publication of Nicholas Georgescu-Roegen's monumental treatise, *The Entropy Law and the Economic Process*.[1] Whereas this book comprehensively discusses the relevant physical theories—thermodynamics and statistical mechanics—its self-confessed objective is to explore the proposition as to whether *similar* theories can be expected to be developed in the realm of economics. The answer seems to be no. It does, however, clearly recognize, and emphasize, the dissipative (i.e., entropic) character of the economic process, thus departing from classical theory. In the interest of completeness, I should mention one other ambitious attempt to introduce materials and energy flow considerations both to ecology and to economics. I refer to H. T. Odum's book *Environment, Power and Society*.[2] As regards ecology, the book appears to be a useful one, but it makes no substantive contribution to economics as a discipline.

make a start on organizing it, suggesting some principles, and outlining some key problem areas. Note that the discussion in Section 3.3 leads directly into the formulation of the materials–process–product model, discussed in Chapter 5. The concluding section (3.4) discusses the need for a new and more rigorous "paradigm" for economics, in which the transformations of materials and energy are conceptually distinguished from the generation of services contributing to welfare.

3.1 IMPLICATION OF THE FIRST LAW OF THERMODYNAMICS

The "first law" states that *matter (mass/energy) is neither created nor destroyed by any physical process.* Until the discovery of radioactivity and Einstein's equivalency law $E = MC^2$, it was assumed that the law of conservation of mass and energy applied *separately* to each chemical species, and to energy. This is now known to be strictly incorrect, since mass can be converted into energy by fusion or by fission.[†] The inverse process is also possible, as exemplified by the creation of an electron–positron pair or a nucleon–antinucleon pair. However, disregarding these exceptional cases,[‡] the first law does hold separately for each species, as well as to aggregates.

At the aggregated level, the first law has already been applied by the author and others[3,4] to the theory of environmental externalities (as discussed briefly in the last chapter). In particular, the so-called materials/energy balance principle states that the sum total of materials and energy extracted from the natural environment as raw materials must exactly balance the sum total of materials and energy returned to the environment as waste flows,[§] less any

accumulation in the form of capital stocks and products inventories. Since many environmental externalities are essentially attributable to the dispersion of waste residuals from production or consumption, one immediate implication is that such externalities are necessarily *pervasive*, rather than *exceptional*, as earlier economic theorists had generally assumed.[†]

The formulation of a materials/energy balance principle in economic language has also stimulated the development of formal links between the theories of optimal growth, optimal exploitation of exhaustible resources, and optimal environmental management. In particular, Mäler's model begins from a formal statement of the materials balance concept (though curiously, the conservation laws are not rigorously maintained in his model). These issues are discussed in the previous chapter.

An obvious point, already alluded to, is that the materials/energy balance principle must also apply to individual economic entities (sectors, regions, firms, activities, households, etc.), and this equality between inputs and outputs must hold true for each chemical *species*—and energy—as well as for aggregates. The foregoing statement may seem trivial —and it is, from one point of view—but the consequences for the economic theory of the firm are not quite. We consider this issue next in greater detail.

Production Functions for Materials-Processing Activities

In economics, a "firm" is any entity organized to produce a product or service that is (or could be) sold in a market. (It need not be autonomous as regards management or ownership.[‡]) Each such entity has a *production function*, depending on its

---

[†]Nuclear explosives, nuclear power plants, and solar radiation depend alike on this interconversion possibility. Even the alchemists' dream of producing gold from "base" metal is now possible (though much too expensive to be of any practical interest).
[‡]Exceptional on the surface of the Earth, that is.
[§]One significant qualification to this statement deserves restatement here, since it is commonly ignored by theorists: The total quantity of waste flows actually *exceeds* the quantity of materials extracted by explicit mining, drilling, or quarrying processes by virtue of the fact that many physical processes utilize (and pollute) air or water that are obtained as free goods. Moreover, some waste products are generated entirely by processes that occur beyond the productive activity, entirely within the environmental media. Combustion processes, in particular, depend upon atmospheric oxygen, which combines with hydrogen or carbon in fuels to form H_2O or CO_2 (as well as CO). If excess air is present in the combustion process some of the atmospheric nitrogen will also be oxidized to form NO or NO_2. The materials/energy

balance in such processes thus requires inclusion of nonpurchased inputs. As an example of pollutants generated entirely by "post-economic" processes, we might consider the formation of reactive oxidants ("smog") in the atmosphere or the formation of sulfuric acid droplets from SO_2 in the atmosphere or in acid mine drainage. The materials/energy balance, for a sector or firm, must, therefore, be computed on the basis of a suitably broad definition of economic activities, extended to encompass posteconomic (delayed or physically remote) reactions that may convert harmless materials into pollutants. Or else one must be content to exclude the products of such reactions from the accounts.
[†]This perspective is emphasized in References 3 and 4. The theory of externalities per se has, however, been discussed by a number of economists, notably Scitovsky,[6] Coase,[7] Buchanan and Tullock,[8] Davis and Whinston,[9] Turvey,[10] and Baumol.[11]
[‡]Thus a firm corresponds closely to an "establishment" as defined by the U. S. Census of Manufactures for purposes of gathering economic data on production, employment, and the like.

specific capabilities, that is a mapping between factor inputs and outputs and that describes the *maximum output* that can be obtained for any combination of inputs. The production function is price-indepenent, by definition: It is supposed to represent strictly *technological* relationships. A profit-maximizing firm (presumably) "knows" this function and uses it to determine the optimum output (quantity and quality) in relation to its markets and in relation to input requirements.

Underlying the basic concept of a production function—and absolutely essential to it, is the notion of *substitutability*. That is to say, it is presumed that a given product (or, a given set of joint products) can be obtained from a variety of different combinations of inputs, and that one such combination will turn out to be "optimum" under any given objective function and set of factor prices. Economists have given less attention, however, to the physical constraints on the possibility of free substitution between fundamentally different factors. This is not to say that the abstract possibility of such constraints or limitations has gone unnoticed, or that the implications have not been explored to some extent.[†] The point that seems worth making here is, simply, that such constraints are real and pervasive, arising as they do from the fact that production is intimately concerned with physical attributes of real materials, machines, and chemical processes.

To clarify the issues, it is helpful to distinguish inputs according to whether they are *complementary (synergistic)*, *competitive (antagonistic)*, *or fixed*.[‡] As an example of complementarity or synergy, take the relationship between tractors and gasoline in agriculture. They are linked together: Neither is of any use without the other. Hence there is absolutely no possibility of substitution between them. (More gasoline cannot reduce the need for tractors or vice versa.) Agriculture also offers numerous examples of competition, such as the relationship between tractors and mules. A farmer might need one or the other,

but probably not both. Or a slightly less obvious example might be irrigation versus land. Irrigated agriculture requires *less* land, hence the two are competitive. Finally, sunlight and water (from some source) are *fixed* inputs—per unit of output—in agriculture, in the sense that nothing can substitute for them.[†]

Table 3.1 shows some of the key relationships for other (aggregate) sectors of the economy. It is clear from Table 3.1 that the neoclassical concept of a production function is reasonably in accord with the actual state of affairs in the *extraction* sectors and in the *service* sectors. That is to say, either labor or capital can be substituted for the other more or less without a priori limit. On the other hand, a large and important area of the economy is concerned with processing and/or fabricating materials or material components. In these sectors, material inputs bear a fixed (or nearly fixed) relationship with material outputs, and while labor and capital can largely substitute for each other, they cannot significantly reduce requirements for materials or energy that are "embodied" in the product; nor can increasing material inputs reduce the need for capital or labor inputs.

Like all generalizations, this one has important caveats. There are a few postextractive processing industries,[‡] in particular, where waste products are comparable in mass to useful products. Under these circumstances, greater investment of capital and/or labor *can*, sometimes, substantially reduce material inputs (or, equivalently, increase saleable outputs by utilizing potential wastes). Lumber and pulp mills, canneries, and meat packers are good illustrations in point. Further, it must be recognized that certain material inputs—notably fuels, refrigerants and coolants, lubricants, cleaning agents, packaging materials, solvents and "propellants"—are not embodied in the (material) product but, rather, become intermediate wastes at the point of use. In many cases, these materials are, in effect, substitutes for labor or capital. (For instance, an acid may be used to clean the surface of a slab of metal, though, in principle, the job could be done by means of a cold chisel and wire brush or a grinding wheel. Or a

---

[†]An extremely sophisticated and highly mathematical theory of production functions has been developed in recent years by Carlson,[12] David,[13] Jacobsen,[14] Shephard,[15] Smith[16] (and many others), which can hardly be summarized here. Suffice it to say that few mathematical stones have been left unturned.

[‡]The difference between competitive inputs and fixed inputs is, more precisely, as follows. Competitive inputs are such that the degree of possible substitution has no a priori technological limit: In principle, one factor may completely replace the other (though it may never do so in practice). "Fixed" inputs are such that application of other factors may only eliminate marginal wastage: There is a definite physically determined lower limit that can only be approached asymptotically.

[†]A distinction must be made between irrigation and water. The results of irrigation increase water availability *per unit of land area*. Irrigation is not necessary as such for growing corps. But water from some source is, obviously, an absolute requirement.

[‡]Combustion wastes obviously constitute a special case, to be discussed separately. Energy conversion (including electric power generation) is thus excluded from the processes discussed in the succeeding paragraphs.

Table 3.1 Characteristics of Factor Inputs by Sector

Factor relationship	Aggregate sectors			
	Extraction of renewable resources (mainly agriculture)	Extraction of nonrenewable resources (mining, drilling)	Manufacturing and construction	Transport distribution and services
Complementary inputs		Machines and structures Fuel and maintenance Utility services and "infrastructure" Nonembodied materials		
Competitive inputs	Above versus labor		Above versus labor	
	Land versus labor Land versus irrigation Land versus fertilizer			
Fixed* inputs	Sunlight Water	Materials embodied in output		None

*As noted already, this implies a limited range of substitution of other inputs to reduce wastage, but a definite lower limit.

volatile solvent may be used to permit a coating to be sprayed on a surface, even though in principle it could be applied by hand. Thus it is important to distinguish *embodied* materials from *labor-saving* materials. The latter (including fuel) is discussed again later. The remarks immediately following, however, refer mainly to sectors, such as metallurgy, chemicals, stone, glass, and clay products, as well as construction in which most nonfuel material inputs (and a significant amount of the energy inputs) are embodied in products.

One is led, in short, to conclude that neoclassical production functions are basically inappropriate for the materials-processing sectors of the economy. More specifically, they effectively violate thermodynamic principles, by implicitly permitting labor or capital to "create" (or "destroy") material or energy inputs.

It is illuminating to list some properties that a materials–process (MP) production function "should" have and compare these with the properties of well-known neoclassical production functions, as illustrated in Table 3.2. (The list is not necessarily complete.)

Returns-to-Scale

Material inputs must increase almost in strict proportion to primary material outputs. Because of moderately increased conversion efficiencies or possibilities of controlling or recycling in-plant wastes in a large plant vis-à-vis a small one, returns-to-scale will normally be slightly larger than unity; that is, a unit increase in output requires slightly less than a unit increase in materials or energy inputs.

Labor and/or capital inputs typically increase much less rapidly than output, as scale increases. For example, a common engineering rule of thumb for capital inputs is the 0.6 power of output. Hence returns-to-scale for capital/labor tend to be substantially greater than unity in the normal range. However, as productive capacity becomes very large the marginal utility of further increases in size will gradually disappear. In other words, returns-to-scale should continue to increase, but at a decreasing rate. At infinite output, a unit of additional output will require an exactly proportional increment of additional input.

Elasticity of Output

In a typical industrial activity, we expect a unit addition of labor or capital—in the *absence* of a unit of additional materials—to yield very little additional output at the margin, whereas a unit of material inputs would yield roughly one unit of output. The sum of all three partial elasticities for labor, capital, and materials must, of course, be

Table 3.2 Comparison of Production Functions

Production function type	Linear	Cobb–Douglas	Constant elasticity of substitution (CES)	Input–output	Activity analysis
Formula for q (2 inputs)	$a_1 x_1 + a_2 x_2$	$b_0 x_1^{b_1} x_2^{b_2}$	$c_0[c_1 x_1^{-\beta} + c_2 x_2^{-\beta}]^{-\frac{h}{\beta}}$	$\min\left\{\dfrac{x_1}{d_1},\ \dfrac{x_2}{d_2}\right\}$	$\sum\limits_{k=1}^{p} k^{y_k}$ where
Parameters	a_j = marginal product of ith input	b_0 = scale factor b_j = elasticity of output with respect to j^{th} input	c_0 = scale factor c_j = distribution parameter h = degree of homogeneity β = substitution parameter	d_j = amount of $j + n$ input needed to produce one unit of output	$\sum\limits_{k=1}^{p} jk\,y_k \leq X$; p = number of activities y_k = level of kth activities e_{jk} = input of jth input to run kth activity at unit intensity
Marginal product $MP_j = \dfrac{\partial q}{\partial x_j}$	a_j	$\dfrac{b_i}{x_j}\,q$	$MP_j = \dfrac{h c_j x_j^{-\beta-1} q}{[C_1 x_1^{-\beta} + C_2 x_2^{-\beta}]}$	N.A.	N.A.
Price ratio at maximum output $P_1/P_2 = MP_1/MP_2$	$\dfrac{a_1}{a_2}$	$\dfrac{b_1}{b_2}\left(\dfrac{x_2}{x_1}\right)$	$\dfrac{c_1}{c_2}\left(\dfrac{x_1}{x_2}\right)^{-\beta-1}$	N.A.	N.A.
Elasticity of substitution $\sigma_{12} = \dfrac{-d\ln(x_1/x_2)}{d\ln(MP_1/MP_2)}$	∞	1	$\dfrac{1}{1+\beta}$	0	0
Elasticity of output $\varepsilon = \sum\limits_{j=1}^{n} ej = \sum\limits_{j} \dfrac{x_j MP_j}{q}$	1	$b_1 + b_2$	h	1 if $\dfrac{x_1}{d_1} = \dfrac{x_2}{d_2}$	1

Source: Reference 17 (adapted).

greater than unity because of returns-to-scale; that is, increasing *all* inputs simply increases the scale of the operation.

Elasticity of Substitution

As already noted, elasticity of substitution between labor and capital may be quite significant. In fact this is normal. But elasticities between labor and materials would be almost vanishingly small in most industries since neither labor nor capital can *create* materials—they can only reduce wastage in the process of transforming them.

Isoquants

A number of different types of production functions are compared with the materials–process (MP) function in Figure 3.1, assuming intersection at one particular combination of inputs (X_1, X_2), where X_1 is a material resource input and X_2 is some aggregate of labor and/or capital. The materials–process case describes a process in which some resource input (X_1) is converted into a product by some aggregate combination of labor and capital (X_2). The shape of the production function shows that not less than $\overline{X}_1$ will suffice to produce the output *regardless of how much labor/capital is expended*. This is a simple consequence of the fact that the material of which

the product is physically composed cannot be *created* by the process. By the same token, at least $\overline{X}_2$ labor/capital will be required to produce the unit of output regardless of the amount of material input supplied. An excess of input material will only increase the labor/capital requirements attributable to materials handling, sorting, storage, and so forth. The optimum combination of inputs (for some combination of prices) is $(\overline{\overline{X}}_1, \overline{\overline{X}}_2)$. The difference $\Delta X_1 = \overline{\overline{X}}_1 - \overline{X}_1$ is the maximum potential saving of material inputs that can be achieved by lavish use of labor/capital, whereas the difference $\Delta X_2 = \overline{\overline{X}}_2 - \overline{X}_2$ is the maximum saving in labor/capital that could be achieved by lavish use of material inputs. If the materials were lavishly supplied, most of them would be wasted, of course.

It is clear that in the narrow region around the optimum point, a Cobb–Douglas (or similar) production function is qualitatively more satisfactory than most of the others, but it has the wrong asymptotic properties. The best asymptotic approximation would appear to be the activity–analysis production function.

Total and Marginal Products

As already noted, the materials-process production function is one in which returns-to-scale are increasing, but at a decreasing rate, as increasing

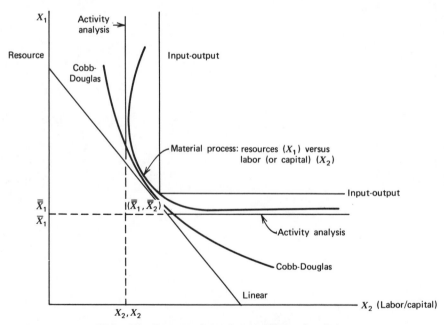

Figure 3.1. Isoquants for various production functions.

labor/capital are applied. Thus the total output as a function of labor/capital approaches a maximum asymptotically, and marginal productivity declines from its peak value but never actually reaches zero. However, returns-to-scale do *not* increase indefinitely as more material inputs are supplied for a fixed amount of labor/capital, as Figure 3.1 shows. On the contrary, beyond some point, *added inputs will have negative utility*. In this case, there is a definite maximum level of output corresponding to finite inputs, and marginal productivity should decrease to zero at the same point, as shown in Figure 3.2. None of the neoclassical functions behave in this way, unfortunately: They all have marginal products that approach zero asymptotically.

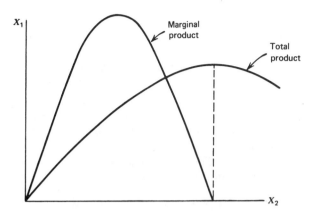

Figure 3.2. Marginal and total product curves.

"First Law" Concepts of Efficiency

Before concluding the discussion of implications of the first law of thermodynamics, it seems worthwhile to mention an application where the "first law" is misleading and leads to erroneous results.

We begin by discussing the concept of energy efficiency as it applies to energy conversion and energy/materials transformations. This topic has received a good deal of attention in the past several years,[18–20] and it appears sufficient to summarize the major points without detailed justification. A list of possible measures of efficiency is given in Table 3.3, which follows. In reading the table, it is important to note that we have normally equated "useful work" with either mechanical energy (e.g., rotary motion) or electric energy. Of course, in the case of an electric motor or generator, the input energy is in one of these two forms and the output is in the other, and interconversion does involve some losses.

Table 3.3   Figures-of-Merit for Various Thermodynamic Processes

Measure		Example(s)	Asymptotic limit
(1)	$\dfrac{\text{Work out}}{\text{Work in}}$	Electric motor or generator torque converter	1.0
(2)	$\dfrac{\text{Work out}}{\text{Enthalpy in}}$	Heat engine	$1 - T_A/T_H$
(3)	$\dfrac{\text{Heat transferred}}{\text{Enthalpy in}}$	Burner or boiler	1.0
(4)	$\dfrac{\text{Heat added to reservoir}}{\text{Work in}}$	Heat pump	$[1 - T_A/T_R]^{-1}$
(5)	$\dfrac{\text{Heat removed from reservoir}}{\text{Work in}}$	Refrigerator	$[T_A/T_C - 1]^{-1}$
(6)	$\dfrac{\text{Available work out}}{\text{Available work in}}$	Chemical or electrochemical process	1.0

where T_A = Ambient Temperature
        T_R = Receptor (warm) Reservoir Temperature
        T_H = Hot Reservoir Temperature
        T_C = Cold Reservoir Temperature

Source: Reference 21.

It may be noted, further, that all of the thermal energy in a reservoir at infinite temperature can be converted into work, but when the reservoir temperature is finite, some of the heat energy is unavailable *in principle*. Entropy is the measure of this unavailability. Thus the temperature at which heat energy is produced or used is a critical quantity in determining the efficiency with which it can be used.

The most widely used figures-of-merit are numbered 1–3 on Table 3.3. The general form that encompasses all three cases follows:

$$\eta = \frac{\text{energy transfer (of a desired kind)}}{\text{total energy input}}$$

This ratio has been called "first law" energy efficiency,[20] since it measures the total energy transfer from the initial state to a final state. Cases 4 and 5 are called "coefficients of performance," since they have values greater than unity. These measures are specialized to heat transfer processes and are not true efficiencies.

The efficiency measure denoted 6 on Table 3.3 is discussed in the next section (3.2.4), since it involves an energy quantity that is *not* conserved by all processes and is thus more properly a "second law" concept. Notice that 6 is equivalent to 1 for those processes where 1 is applicable, but it is not identical to 1. It is also interesting to observe that, except for 1, all "figures-of-merit" are different from unity *only* because numerator and denominator refer to different measures. If all energy inputs and outputs (including flows to and from the environment) were taken into account in each case, the efficiency would be identically unity (i.e., input = output), which is not a very useful result.

3.2 IMPLICATIONS OF THE SECOND LAW OF THERMODYNAMICS: ENTROPY

Entropy is one of the most mysterious and opaque of all the concepts of modern physical science. It is a measure of the extent to which the energy contained in a system is available for purposes of doing useful work on some other system. If ΔS is an increment of entropy and T is the absolute temperature, the product $T \Delta S$ represents energy that becomes unavailable as a result of a process. The famous second law of thermodynamics states that for *any* process within a closed system—when all contributory factors are taken into account—the entropy of unavailable energy associated with the state-of-the-system always increases.[†] This unidirectional characteristic has tempted some philosophers of science to identify the forward direction in time as the direction in which entropy increases. Thus entropy becomes "time's arrow." Without it, all physical laws would be completely symmetrical with respect to the direction of time.[‡]

Entropy has been given an alternative definition in "statistical mechanics" (by Boltzmann, Gibbs, Maxwell et al.) namely, as the logarithm of the probability of random occurrence of a given "state" of matter, as measured in terms of both positions (or locations) and energy levels (or momenta), of particles or components of the system.

From an engineering point of view, the meaning of the second law is, simply, that the ability of a thermal energy-conversion system to convert heat energy into kinetic energy depends upon the existence of large temperature gradients in the system. In other words, the heat engine must have both *hot* and *cold* parts in order to work. Moreover, every time some of the heat is converted into work, the hot–cold differential is decreased: The hot parts of the engine are cooled and the cold parts are warmed up. Nonthermal energy conversion processes also require the existence of field gradients, for example, of voltage or gravity. Converting chemical or gravitational potential energy into work involves reducing the state differentials between the parts of the system.

The second law implies that all physical or chemical activity tends to promote this process of approaching temperature equalization or "thermal equilibrium." When thermal equilibrium in the universe is ultimately achieved, no more directed activity is possible: All motion is random. This ultimate state has sometimes been called the "heat death" of the universe. Statistically speaking, the state of ther-

---

[†]However, it must be noted, there are some processes where the entropy of a *subsystem* can decrease temporarily at the expense of some larger system.

[‡]It follows that if any cosmological process is capable of reversing the direction of time—even "locally"—it must also reverse the buildup of entropy. This point is interesting in view of the fact that some conceivable distributions of ultradense matter lead to equations of motion in which time change *is* actually reversed. Some relativistic theorists are now seriously exploring the implications of these curious mathematical results. If any of this is confirmed by experiments, we would have to conclude that the second law of thermodynamics is a local—not a universal—phenomenon.

mal equilibrium is the state that is the "most probable," most random and disordered. A state of low entropy intrinsically has a high order and low probability.

The ultimate state of heat death is approaching very slowly, however (if at all), in terms of the time scales of interest to man. As long as the surface temperature of the sun is in the neighborhood of 5000°K—and will remain so for hundreds of millions, if not billions, of years to come—and the temperature of interstellar space remains at close to 0°K, it will continue to be possible to convert this temperature gradient into useful work.[†] The existence of self-organizing biological systems on Earth, capable of maintaining themselves (i.e., their structural parts) in states of very low entropy, is directly attributable to the Earth's location in the thermal gradient of the sun.

This does not altogether eliminate the problem of entropy increase as a significant one for man, however. Inasmuch as entropy is a measure of the intrinsic probability of a state, it can be correlated in some cases with the vaguer and looser concept of *disorder*. With few exceptions, entropy increases as disorder increases. The primary example of a low entropy state is one in which there are two distinct regions: one of almost infinitely high temperature and one of nearly zero temperature.[‡] The orderli-

ness is implicit in the separation between the regions.

For future reference, it is evidently important to distinguish between energy per se and energy actually available for doing work. In effect, energy comes in various "quality" levels, depending on how much work can be extracted from it. Electricity is, in effect, energy of the highest possible quality, since it is virtually 100% "available work." Thus we can roughly judge the quality of a nonelectric energy source by the amount of electricity that can be obtained (in principle) from it. In effect, electricity is the most "orderly" form of energy because it arises from a complete separation of positively and negatively charged particles. This state is obviously very improbable in nature.

Of course, the concept of physical order extends beyond the separation of hot and cold regions, or positively and negatively charged particles. It can be generalized, also, to the separation of physical or chemical elements of different species. A large deposit of some metal such as silver in pure form represents available energy—and a kind of physical order. When this material is mined, converted into various products such as photographic film, and eventually dispersed widely (and, essentially, irrecoverably), the original available energy is lost, and order has become disorder. Entropy has increased.

Entropy and Order

It is convenient, at this point, to introduce the concept of *negentropy* (intrinsic order) even though this relationship is essentially metaphysical, not rigorous. By extension of the foregoing remarks, it is clear that high quality (i.e., pure or nearly pure) natural resources constitute a kind of *stock of negentropy*. As the highest quality resources are used up (and dispersed in low quality form), the available stock of negentropy is decreased.

It is true, of course, that low quality[†] natural resources, such as copper or silver ore, can be upgraded or beneficiated by various means. But all of

[†]In the very long run, be it noted, there is no reason energy cannot be collected in other parts of the solar system—especially the moon—and brought back to the surface of the Earth. The very large temperature differential between the moon's "hot" and "cold" sides—maintained by radiation *from* the sun on the one side and radiation *to* interstellar space on the other—constitutes an enormous potential energy resource which could, in principle, be exploited by man. A variety of possible heat engines to utilize this temperature difference has been suggested by physicists. The problem is how to get the stored energy most efficiently back to Earth. Again, a variety of technical means has been suggested, from direct transmission via microwave beams and lasers to using the moon as an industrial site and shipping finished consumer products or synthetic foods back to the Earth. A very important constraint must be noted, however; the use of the moon as a supplementary energy source for the Earth must *not* be permitted to raise the average temperature of the surface of the Earth significantly, or disastrous and irreversible climatological consequences could ensue. The relevant balancing factors are somewhat too complex to discuss fully in this footnote, but they include the CO_2 level of the atmosphere, the water vapor cycle and cloud cover, the amount of snow cover, the state of the ocean surfaces, and the density of very small particles in the stratosphere.

[‡]The universe was presumably like this at the instant of the "Big Bang" from which (cosmologists now think) it all began!

[†]The notion of "quality" as applied to resources is not unambiguous. Typically, resource economists or mining engineers measure quality in terms of the percentage of metal in the ore (e.g., 0.5% copper ore). However, this measure emphatically does *not* apply to different ore types. A scrap automobile weighing 4000 lb containing, say 40 lb (1%) of copper is *not* higher quality, regarded as an "ore." The appropriate measure of quality is, really, the difficulty of separating the pure material from its existing form.

these methods of upgrading depend on the use of energy. The greater the level of upgrading (concentration) required, the more available work is needed. Technological advances may increase the efficiency of such processes, but there is always an irreducible minimum energy requirement for doing so. Thus the negentropy of physical materials can be restored by the appropriate application of high quality energy (i.e., available work). But at present, this energy itself comes largely from accumulated fossil sources that are thereby depleted. Moreover, the negentropy depletion from using petroleum or natural gas resources to generate electricity to separate aluminum from bauxite (for instance) far exceeds the negentropy *addition* to the purified aluminum.

Clearly, the faster we use up high quality nonenergy resources (depleting one type of negentropy stock), the more we accelerate the demand for fossil energy resources, thus depleting another type of negentropy stock. As the better quality fossil energy resources (such as petroleum) approach exhaustion, a further increment of demand acceleration is added since the remaining fossil energy fuels, such as coal, are of still lower quality. (More energy is needed to extract and refine coal to a given level of quality than is needed for petroleum, for instance.)

The extent of the depletion of negentropy in the form of high quality resources, can be explored in another way by noting that the natural geophysical processes that create ores can be regarded as generating some distribution of levels of concentrations. This distribution will generally have a maximum value at some concentration close to the average concentration of the resource in the substrate (see Figure 3.3). That is to say, out of 100 random samples of the substrate, ranked by increasing concentration, only a few would reflect either the upper or lower extremes of concentration. Most would be somewhere near the average[†]—a little above or a little below.[‡] However, the *recoverable* portion would as a rule be only that fraction of the resource located in the most concentrated deposits.

It is clear that many—if not most—resources are indeed distributed in such a manner.[22] It is also clear that the distribution function is a direct measure of the intrinsic probability of a state (of concentration).

[†]The average might be taken with respect to a major subcategory such as crustal rock, sedimentary rock, or sea water, for instance.
[‡]A "normal" or "log-normal" distribution would have these characteristics, of course. However, the real distribution need not be normal in the least.

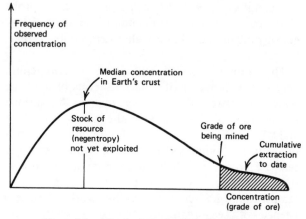

Figure 3.3. Distribution of exhaustible resources.

As noted earlier, the associated (negative) entropy associated with a state is the logarithm of its probability of random occurrence. Hence the rate of negentropy depletion per unit time is roughly proportional to the time rate of change of the logarithm of the concentration of the one currently being extrated, namely,

$$- \dot{S} \approx \frac{\partial \ln C}{\partial t} \qquad (3.1)$$

where S is entropy and C is concentration.[†]

In short, the depletion of negentropy itself increases the demand for exhaustible resources, quite apart from the level of final consumption. In other words, even if final consumption of material goods remains constant—and extraction/processing technology were also unchanging—the rate of extraction will necessarily increase, at an increasing rate! In the absence of technological improvements (including, but not limited to increased recycling), a "crash" must necessarily ensue even if there is no increase in population or per capita consumption. (Where both of these are also increasing, the inevitable crash will simply come sooner.)

It is interesting to note that the classical economic theory of exhaustible resources, as developed by Hotelling, Herfendahl, and others, does predict the exploitation of the highest quality (i.e., lowest cost)

[†]It is interesting to speculate on the conditions under which $-\dot{S}$ itself will increase. Qualitatively, it is clear that it depends on the *shape* of the ore distribution function, that is, the rate of increase of the amount of ore available at decreasing concentrations. For most ores, the distribution is apparently such that the decline of marginal concentration will slowly *decelerate* with time. However, there may be exceptions to this rule.[23]

reserves first,[†] *but has not, so far, allowed for the positive feedback between decreasing quality (negentropy) of the remaining resources and the rate of extraction.* Nor have the increasing environmental costs of this entropic buildup been incorporated in models of optimal extraction.

Another important point deserves to be considered. It is, simply, that the eventual dispersion of some kinds of exhaustible natural resources as low grade wastes results in degradation of the natural capital which generates so-called renewable resources. This can already be seen in a number of ways—quantitatively minor, so far, but qualitatively ominous. Loss of vegetation cover, increased surface temperature, erosion, and loss of topsoil (combined with silting of streams, raising of river beds, and increased flood hazard) probably constitutes the most important single mechanism of environmental degradation, so far. In the temperate zones, the major causes are "clear-cutting" of timber, excessive cultivation (especially of hillsides), plus mining, construction, and road building activities. In the Mediterranean basin and the tropics, the chief villain is overgrazing by goats, sheep, and (in India) cattle. Overpopulation is the ultimate problem in the less-developed part of the world. Excessive zeal for short-term extractive profits was the chief problem in the more developed countries, but public sector investment in reforestation and erosion control, combined with education and improved agricultural/forestry practice has now somewhat ameliorated the problem.

Pollution, resulting from flows of wastes back to environmental reservoirs, may be overtaking erosion as the chief cause of degradation of natural capital assets. Eutrophication—actually, overfertilization—is severely reducing the value to Man of many fresh water lakes and streams. Oil spills and build-ups of DDT, PCBs, and other long-lived toxic residuals in the oceans are beginning to pose a significant potential threat to marine food chains. (While the loss is probably not yet very large, the state of the oceans is so nearly irreversible that there is little chance of any subsequent amelioration.[24] Nor do we know, yet, how long the lags are between an initial perturbation of the system and its approach to a new equilibrium state. Damages observed to date may be merely preliminary harbingers.) Some raptor species of birds are threatened by DDT buildup in their food chains. Heavy metal discharges have already caused catastrophic (if localized) epidemics in Japan[†]—with many dead and hundreds or even thousands permanently disabled—and the morbidity rate shows little sign of abatement, even though the offending discharges were stopped more than a decade ago. Heavy use of toxic chemical pesticides is also a subtler (but no less damaging) form of environmental degradation, since harmless species and helpful natural controls tend to be preferentially destroyed. Thus the natural ability of the ecosystem to ward off an outbreak of pests is interfered with. Similarly, use of chemical fertilizers harms the natural soil organisms that condition the soil and permit plants to extract nutrients from inorganic soil materials. If chemical fertilization is subsequently withdrawn, the soil tends to be less productive than it was previously. Numerous other examples can easily be cited.

From a physical science viewpoint, this degradation can also be equated with *negentropy depletion*. In effect, the various ways in which human activity disrupts and interferes with natural processes is a kind of externally imposed loss of order. While agriculture may appear to embody a greater degree of order than a natural ecosystem, this is entirely misleading. The agricultural system involves only a few species operating in an artificial state of controlled disequilibrium with the help of machines, fertilizers, pesticides, hybrid seed, and so forth. All of this activity receives a large negentropy subsidy from other parts of the industrial system. If these inputs were withdrawn, the system would go into violent unstable oscillations until, presumably, natural stability returned as untamed species from the surviving natural ecosystems reinvaded.

As mentioned in Chapter 1, the tendency of natural systems to be stable is a function of the large number of different species that are present. It is precisely this differentiation that gives the system a

---

[†]A contrary case has been derived by Mäler (Chapter 2 op. cit.) under highly restrictive and unrealistic assumptions, namely, a zero discount rate over a finite planning period, together with a decreasing rate of consumption and increasing capital costs. Even if this scenario made any sense, the optimal solution is totally inconsistent and infeasible in practice. If one does not exploit the best resources first, by implication, one should begin with the "worst." This is a mind-boggling thought. Do we start by extracting gold from seawater, leaving the South African mines as a reserve for use in the future when capital is more expensive? But no, surely there is still a lower quality source to begin with. Why not mine the moon first?

[†]Minimata disease (from mercury ingestion) and Itai-Itai disease (from cadmium ingestion).

high degree of negentropy or order. The more species are present, the more intrinsically improbable the system is. As species differentiation is reduced, entropy increases as the system becomes more "probable."[†]

From the economists' point of view, the key point that has been made is that *dispersion of waste products into the environment*—as a result of the extraction and use of exhaustible resources—*causes long-run reduction in both potential productivity and stability of the natural capital.* Reduced natural productivity can be overcome to some extent by external intervention—as exemplified by the use of fertilizers and hybrid seeds—but not without further depleting the exhaustible resources that (at the moment) are sustaining our economy. The same is true of stability. As natural stability is degraded as a result of human intervention, an artificial stability can be imposed externally, to some extent. Again, the use of chemical pesticides or the building of dams and dikes to counter the threat of floods can be cited as instances.

But this artificial stability is not costless, and it requires the expenditure of "available work" from fossil or other energy sources. This is another example of a feedback phenomenon that results in accelerating the rate of exhaustible resources. The sequence is this: An extraction occurs; waste discharges eventually follow. These reduce the productivity and the stability of the natural environment. To compensate, external intervention is required. But intervention itself constitutes a demand for additional exhaustible resources. This feedback loop has not been incorporated into recent economic models of the utilization of exhaustible resources.

Technology and Negentropy

Technology has in recent years become a kind of *Deus ex Machina* for economics, especially since the recent "Limits to Growth" controversy.[‡] Technologi-

cal change is increasingly relied on as the "escape hatch" from neo-Malthusian dilemmas which otherwise seem very formidable. For example, R. M. Solow said in the prestigious Ely lecture (1974),

If it is easy to substitute other factors for (exhaustible) natural resources, then there is, in principle, "no problem". The world can, in effect, get along without natural resources. Exhaustion is an event, not a catastrophe. ... If, on the other hand, real output per unit of resources is effectively bounded—cannot exceed some upper limit of productivity which is in turn not too far from where we are now—then catastrophe is unavoidable. ... Fortunately, what little evidence there is suggests that there is quite a lot of substitutibility between exhaustible resources and renewable or reproducible resources.

In general, however, economists have preferred not to examine the phenomenon of technological substitution too closely. Rather, elasticity of substitution is typically regarded as an intrinsic "property" of supply and demand functions, which in turn are generally specified for mathematical convenience (or not at all).

Thus in classical economics, if a temporary imbalance occurs between supply and demand, the problem will be resolved by a price adjustment that tends to decrease the demand and increase the supply (or conversely). On the supply side, a small enough adjustment may involve simply adding more labor to increase output,[†] whereas on the demand side, final consumption of one product may decrease slightly with no compensating increases elsewhere.

As the magnitude of the price adjustment increases, however, the scope of the adjustment process becomes more comprehensive. Eventually, as the price rises higher, suppliers must begin to phase *in* alternative sources or production processes; whereas consumers, on the other hand, phase *out* marginal applications, shifting over to cheaper substitutes, the demand for which then increases. To the extend that supply elasticity is relied upon to achieve this market-clearing equilibrium, technological change is implicitly assumed. If supplies become generally inelastic at some point (because of technological or other limits), higher prices would simply mean reduced consumption,[‡] hence reduced welfare.

[†]Another way of looking at this question (Chapter 1) is by noting that the *negentropy* has the same properties of *information*.[25, 26] The negentropy level of a natural ecosystem can thus be equated to the information content of its gene pool.[27] Each species embodies in its genes the entire accumulated experience accumulated during its evolutionary development. If a species disappears, this information is irretrievably lost.

[‡]The issues and arguments will *not* be recapitulated here. In case some reader has recently returned from a Trappist Monastery or has recently recovered from total amnesia, he or she might wish to refer to References 28, 29, or 30.

[†]This is true if the output is extractive or if it is a service. (Recall the discussion in Section 3.1.)

[‡]No compensating increases for substitutes being possible by assumption.

It is generally accepted that, up until now, technological improvement has (seemingly) more than compensated for the depletion of natural stocks of negentropy. As noted in the previous chapter (Section 2.2), the only way to reconcile the "Hotelling principle," whereby resource prices should remain the same over time at the real market interest rate, with the declining or nonincreasing price trends observed and analyzed by Barnett and Morse, is to invoke rapid technological change.[†] To the extent that technology has increased our ability to extract low quality resources, it has effectively increased their quantity and quality. In this sense, exploration and extraction technology are virtually equivalent to the resources themselves—hence *technology is effectively a negentropy stock*.[‡]

Actually, the word "virtually" conceals a fairly large caveat, which must be recorded. While exhaustible resource *prices* have been generally, if erratically, decreasing since 1870 (at least, up to 1972 or so), the energy required per unit of extraction has probably substantially increased, despite improvement in the efficiency of machines. Negentropy stock comparisons must take this additiona negentropy input fully into account. If this were done, it is not unlikely that the last century would still show a substantial net decline in negentropy stock, notwithstanding technological improvements during that time. But, it is possible in principle, at least, that technology may result in large, if (at least) temporary, increases in available negentropy stocks. The obvious example confronting us today is the possibility of taming the process of nuclear fuel "breeding" and, ultimately, fusion of deuterium and tritium under controlled conditions.

The deeper and more fundamental issue raised by Solow (and others), however, is the following: Can technological improvements continue to compensate for the depletion of natural resource (negentropy) stocks *without limit*? To assert that technology is so potent is tantamount to a denial of the second law of thermodynamics, *since the assertion implies that there must be no limit to the amount of negentropy that can be created by intellectual activity*. If the second law ultimately turned out to have exceptions, or to be a special case, it will not be the first time in the history of science that a major postulate has collapsed. But for the present, it seems wiser to assume that the second law as now stated is indeed valid and that there are absolute upper limits to the negentropy stocks ultimately available on Earth—or in the Solar System or anywhere else, for that matter—including the contributions of future technology.

More to the point, perhaps, this tentative acceptance of the existence of ultimate limits to the capabilities of technology seems to accord with both intuitive and empirical understanding of the dynamics of technological change. Of course, no general law of technological progress or substitution covering all cases can be given. But, more often than not, the rate of progress accelerates exponentially at first but later reaches a peak and begins to decelerate again as a limit of some kind is approached. A measure of the technological capability over time typically appears as an elongated S-shaped—sometimes called "logistic"—curve, as illustrated in Figure 3.4, in which performance is plotted against cumulative experience. The progress achieved per unit of additional experience is plotted in Figure 3.5.

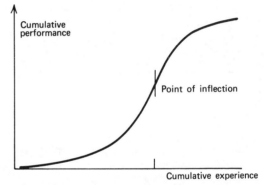

Figure 3.4. Technological progress function.

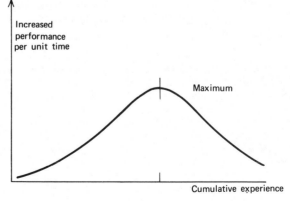

Figure 3.5. Technological progress per unit time.

[†]See references cited in Chapter 2, notably Hotelling, Barnett, and Morse and Schultze.
[‡]This view is fully in accord with modern information theory, which treats "information" as mathematically identical to negentropy.

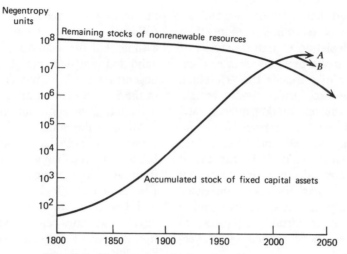

Figure 3.6. Negentropy stocks (assets)—arbitrary units.

Evidently, Figure 3.5 is equivalent to the usual assumption of declining marginal returns to additional investment in a given technology.

There is nothing surprising to economists in the notion of declining marginal returns. It is only the contrary assumption that would be somewhat surprising. It is also somewhat surprising that some people who should know better have adopted the technology gimmick to allow themselves to suppose that *resource output per unit of resource input* might not be effectively bounded.[†]

Negentropy Depletion

During the recent historical period—say 1800–1975 —it is clear qualitatively that while the Earth's stock of recoverable nonrenewable natural resources has been depleted, its stock of "productive capital assets," however these might be defined, has increased. In a sense, nonrenewable resources have been "converted" into assets, including technology. The conversion is not direct, of course. Many nonrenewable resources, especially fossil fuels, have been

physically consumed, whereas many metals mined a century ago are still available for use. However, a machine is much more than the materials of which it is composed. Its labor and technology components are probably much more significant in entropic terms. The situation is depicted graphically in Figure 3.6.

The growing stock of productive capital assets, in turn, have enabled us to (a) increase the actual rate of extraction of nonrenewable resources and (b) increase the efficiency of utilization of potentially available renewable resources, such as fisheries.

The quantitative comparisons between the various curves are not particularly meaningful—certainly the implication that accumulated fixed capital assets will exceed the remaining underground stocks of nonrenewable resources by a specified year (say 2000) should not be taken too seriously. It seems reasonable, however, to suppose that a transitional period or turning point will be reached within a few decades. It would be characterized by several phenomena. In the first place, the rate of extraction of nonrenewable resources, which has been accelerating for two centuries or more, will reach a peak[†] and subsequently begin to fall, as a consequence of approaching exhaustion, leading to rising prices and ultimately falling demand.

Meanwhile, production of conditionally renewable resouces (food, forest products, etc.) has already risen rapidly in the past two centuries, whereas the

[†]Another technological escape hatch to the "limits" bind still remains open, however. It is that final services (as "consumed") may not have an irreducible minimum material or energy content. In other words, it may be technologically possible to increase the "service content" of a given unit of materials or energy effectively without limit. Since service content is nonmaterial, the second law constraints against unlimited creation of negentropy does not apply in this case. This is a much subtler means of escape than the other. For the moment I am unable to think of a conclusive argument against it. However, the question is discussed at greater length in Section 3.3.

[†]It will peak, that is, unless nuclear fusion or successful extraterrestrial energy acquisition ushers in a new era of much cheaper energy. See separate discussion of this point.

upper limits of *potential* renewable production have fallen from earlier levels due to degradation of the environment by erosion, loss of natural fertility and soil condition, increasing ecological instability, and buildup of toxic pollutants. This degradation is attributable to human activities of all kinds, but the high rate of extraction and use of nonrenewable resources is an increasingly important contributing factor. Although it is uncertain how much actual production levels can still be pushed upward by application of advanced technological means, it is highly plausible that *the two curves are converging.* As of 1975, it is likely that actual production of conditionally renewable resources is somewhat less than half but somewhat more than 10% of the maximum feasible level.[†]

The approaching transitional period—or crisis —may lead to either of two qualitatively different outcomes (A, B), depending on how clearly it is foreseen and how well we plan for it. An unfavorable, perhaps catastrophic, outcome (B) will result if a large part of the world's economy 20 or 30 years hence is sill dependent on extracting nonrenewable resources for current consumption. This is an all-too-likely possibility if resource prices are politically controlled at too low a level (to encourage current consumption). Too-low prices for exhaustible resouces will fail to induce the necessary technological substitutions in processes, products, and services. The downward spiral leading to catastrophe could be kicked off if technological inputs to the production of renewable resources (e.g., use of synthetic fertilizers, pesticides, hybrid seed, irrigation water in agriculture) were rather suddenly withdrawn because of an economic crisis. Meanwhile, demands for substitutes for no-longer-available sources of fossil energy would place increased pressure on the productivity of an already heavily stressed agriculture–forestry–fishery system of the Earth. The combination could result in increasing instability and deterioration of the potential productivity of renewable resources and, eventually, a sharp drop in actual production, too. In short, an ecocatastrophe of some sort would be inevitable.[‡]

A more favorable (or, at least, tolerable) outcome (A) would result if—and only if—the production/

consumption system of the Earth begins essentially immediately to get itself "unhooked" from dependence on nonrenewable resources for current consumption. A large fraction of what remains under the ground now will be needed to keep the economic system functioning during the several decades that will, at minimum, be required for a smooth transition to occur. As much as possible of these resources should be reserved for production and maintenance of capital assets; as little as possible should be dissipated in nonrecoverable form. Technological development in the coming decade *must* be focused on this transitional phase, or the crisis will find us unprepared and helpless.[†] It is of overriding importance that scarce natural and capital resources *not* be invested in augmenting large macro-systems of elaborate and interrelated technological elements, some of which are intrinsically (and absolutely) dependent on a continued supply of a nonrenewable resource.[‡] If that resource is exhausted, the entire capital investment in being must be written off—or else a new, and otherwise probably unjustified investment (e.g., coal liquefaction) will be required to keep the system in operation. Since such a macro-system cannot be abandoned in much less than three or four decades, it is essential that the process of substitution be started immediately.

Some possibility does exist that a very dramatic new "backstop" technology can be developed that would change the picture dramatically. Breeder reactors and nuclear fusion have been mentioned most frequently as possibilities. (Indeed, there is still considerable sentiment in ERDA and in Congress for putting most of our eggs in this basket.) In the longer run, extraterrestrial resources may be tapped. However, very powerful technical and environmen-

---

[†]In other words, food production can probably be increased by somewhere between a factor of two and a factor of ten.

[‡]It is probably inevitable anyhow in some overstressed areas, such as Java and Bangla Desh—and, perhaps, all of India.

[†]Beyond the scope of this enquiry, but nonetheless very relevant, is the apparent tendency of human institutions to resist change until the disequilibrium is so great that a violent "shock" occurs, rather than a smooth, well-managed transition. Thus the word *"crisis"* is all too appropriate.

[‡]The automotive and air-transport systems of the world are cases in point. All motor vehicles and aircraft are absolutely dependent on liquid hydrocarbon fuels that must be derived from petroleum. The only alternative aircraft fuel that has seriously been put forward is hydrogen, presumably derived from electrolysis of water using nuclear power as the basic energy source. In the case of motor vehicles on land, the alternative seems to be electric propulsion—again predicated on nuclear power. While these substitutions cannot be ruled out permanently, they could scarcely happen within 50 years from now, even if the necessary nuclear power technology is developed as expeditiously as possible.

tal arguments can be made against depending on any supply technology such as these.[†] But even if all these arguments could be answered, the sheer magnitude of the problem of building enough nuclear power plants to supply all electric power needs alone by the year 2000 or 2010—disregarding the problem of replacing liquid and gaseous hydrocarbon fuel (e.g., by means of a nuclear-based "hydrogen" economy)—is probably beyond our collective means and ability.[38] Other approaches to conserve resources *must* be taken.

Second Law Energy Efficiency[‡]

It will be recalled from an earlier discussion ("First Law" Concepts of Efficiency) that possible first law efficiency measures are often either misleading or trivial. It was pointed out that the only satisfactory efficiency measure for any energy conversion process is 6 from Table 3.2 as follows:

$$\xi = \frac{\text{available work of final outputs}}{\text{available work of inputs}} = \frac{B_f}{B_o} \quad (3.2)$$

Thermodynamically, the available work of an active system is that the maximum amount of useful work B the active system could do on another (passive) system as both systems move to an equilibrium final state with respect to the atmosphere

$$B_o - B_f = E_o - E_f + P_o(V_o - V_f) + T_o(S_o - S_f) \quad (3.3)$$

where E, P, V, T, and S refer to internal energy, pressure, volume, temperature, and entropy, respectively. The useful work W done on the passive systems by the active system is defined by the first law of thermodynamics as the change in enthalpy minus the heat that is lost to the atmosphere and the (nonuseful) work done on the atmosphere. It can be shown that the useful work done by the active system is maximized when there is no entropy change in either the active system or the atmosphere.

[†]A thorough discussion is not appropriate here. Suffice it to say that the "con" arguments that are most frequently cited include the following:
1. Risk of reactor accident,[31, 32]
2. Risk of environmental dispersal of nuclear wastes,[33, 34]
3. Risk of diversion of plutonium to misuse for blackmail or terror purposes by criminals or political fanatics,[35, 36]
4. Risk of accelerated proliferation of nuclear weapons leading eventually to use in warfare.[37]

[‡]This section is largely taken from Reference 21.

This is the case if and only if the process occurs very slowly and reversibly, whence

$$W_{\max} = B_o - B_f \quad (3.4)$$

The available work "in" the initial state is the work required to produce this state reversibly from a system in equilibrium with the ambient atmosphere. Similarly, the available work B_f in the final state is, therefore, equal to the minimum available work $B_{\min}$ needed to produce the final state starting from a system in equilibrium with an ambient atmosphere. These quantities are independent of any specific process. From the foregoing remarks, it can be seen that our definition coincides with that used in Reference 20:

$$\frac{B_f}{B_o} = \frac{B_{\min}}{B_o} \quad (3.5)$$

A list of calculated typical first and second law efficiencies is given in Table 3.4. The industrial process efficiencies given are substantially higher than the energy conversion efficiencies at the top of the table because much of the available work in the input material remains available in the finished material. This is particularly noteworthy in the case of petrochemical-based plastics, where nominal second law efficiencies as high as 80% may be achieved. Obviously, the key to systems efficiency in the use of materials is recycling. If a material once produced is retained perpetually in inventory (as is, for instance, gold), then its available energy is preserved. However, when a material is dissipated or degraded by use to the point of being unrecoverable, then its available energy content is lost. This is an aspect of efficiency of materials utilization, which we discuss next.

One possible point of confusion in Table 3.4 should be cleared up at this point. Most physicists and engineers are accustomed to thinking that energy conversion processes are limited by the so-called Carnot efficiency which, of course, reflects the second law of thermodynamics. But if $B_{\min}$ is limited by the Carnot efficiency, surely present electric generating plants are much closer to their ultimate limits than the 0.33 efficiency shown? The answer is that $B_{\min}$ is not Carnot limited. The reason for this apparent contradiction is the fact that the Carnot limit applies only to a limited class of processes, namely, cyclic (reversible) *heat engines* operating between a high temperature reservoir and a low temperature reservoir. It does not apply to fuel cells, for instance.

Table 3.4 Calculated Efficiencies, Various Processes

Energy conversion process	First law η	Second law ξ
Burner to steam-electric generator	0.33	0.33
Burner to process heat, 1000°F	0.8	0.65
Burner to process steam, 250°F	0.8	0.225
Electric heater to hot water, 120°F	0.75	0.045
Burner to hot water, 120°F	0.50	0.029
Burner to space heat, 70°F (room)	0.60	0.028
Burner to space heat, 110°F (at the register)	0.60	0.074
Burner to space heat, 160°F (at the furnace plenum)	0.75	0.145
Otto cycle engine, constant speed	0.32	0.32
Engine driven heat pump to steam, 250°F	1.18 (COP)	0.299
Engine driven heat pump to space heat, 110°F	2.45 (COP)	0.202
Industrial electric drive	0.90	0.90
Automobile engine to rear wheels, 60 mph	0.14	0.14
Automobile, federal urban driving cycle	0.09	0.09

Materials Transformation Process		
Coke oven	N.A.	0.88
Pig iron, blast furnace	N.A.	0.77
Steel, open hearth furnace	N.A.	0.66
Steel, electric arc furnace	N.A.	0.80
Steel, basic oxygen furnace	N.A.	0.93
Iron and steel finishing	N.A.	0.415
Aluminum, Bayer-Hall process	N.A.	0.45
Aluminum, Alcoa process	N.A.	0.59
Chlorine manufacturing	N.A.	0.40
Ammonia manufacturing	N.A.	0.50
Pulp and paper	N.A.	0.29
Cement	N.A.	0.25
Glass manufacturing	N.A.	0.22

Source: Reference 21.
COP = Coefficient of Performance

3.3 PHYSICAL PRINCIPLES AND THE MICRO-STRUCTURE OF THE ECONOMY

In the next two chapters, this book explicitly addresses the problem of constructing large-scale realistic models to reflect resource supply constraints, from one form to another, and waste residual generation. The purpose of the present section is to introduce some basic taxonomic prerequisites to model building. These can be divided into two groups:

- Resource/commodity stocks and flows.
- Transformations.

Resource/Commodity Definitions

For purposes of this discussion, a *stock* is any quantifiable physical or economic quantity, includ-

ing any of the following:

- Natural systems (arable land, forests, fresh water, lakes and rivers, ocean fisheries, and estuaries).
- Mineral resources (natural gas, petroleum, coal, limestone, and metallic ores).
- Semirefined and scrap materials (pig iron, scrap iron, blister copper, cement, sand and gravel, lumber, and woodpulp).
- Finished materials (metal, castings, strip, sheet, rod, tube, wire and powders; plastic extrusions, moldings, sheet, film, and tube; plywood, paper, plasterboard, and concrete).
- Finished fuels and forms of energy (synthetic gas, gasoline, coke, and electricity).
- Manufactured products.

A *flow* is any shift of a stock from one "reservoir" or category to another. This shift may involve physical transfer between reservoirs or physical/chemical transformation between forms. It may involve only a change in accounting category. For instance, exploration for minerals will increase stocks in the categories of "proved reserves" and "inferred reserves." No new resources are actually created thereby, of course. Similarly, purchases or sales of minerals by a national stock-piling agency changes the ownership category but nothing else. Nevertheless, these changes also constitute flows.

The term transformation has been used loosely, until now. Hereafter, it is helpful to subdivide transformations explicitly among the following major types:

Production of Goods
- Extraction processes (mining, harvesting, logging, and fishing).
- Conversion processes (beneficiation, reduction of ores, alloying, chemical processing, and energy conversion).
- Manufacturing processes (fabrication, assembly, and construction).

Production of Services
- Consumption of material goods and energy (intermediate or final).

The distinctions are clear and self-explanatory in most cases, but some ambiguities may require elaboration. An *extraction* process is one that converts a natural resource into a salable commodity by physical removal. *Conversion* embraces all transformation steps leading up to the production of *finished materials*.

The chief source of ambiguity lies in the definition of the latter. A fairly precise definition could be given, but it would not be useful in practice, since it would often fail to correspond to established sectoral demarcations. To minimize conflict with existing usage, it is necessary to define finished materials in terms of specific industry sector or commodity categories.[†] A list of the input–output sectors and four-digit SIC subsectors is given in the Appendix to Chapter 4. In general, it seems most reasonable to classify industries as conversion if refinement, separation, metallurgical, chemical, or biological

processes are involved. When the processes are primarily physical (e.g., blending, forming, and packaging), the sector is counted as manufacturing. Inconsistencies inevitably crop up. Thus iron and steel castings and forgings are traditionally lumped with basic iron and steel, whereas these processes are separated for nonferrous metals. Weaving is counted as conversion because the product is simply cloth, whereas knitting is counted as manufacturing because the product is generally a garment. The only justification that can really be given is convenience, and the conjecture that it probably does not really matter much exactly where the boundary between conversion and manufacturing is placed.

The production of material goods from finished materials comprises the next stage of transformation. The industrial sectors involved here include all industries (other than extraction and conversion) that produce material products. A list would not be illuminating and can be omitted.

Finally, some industries—as well as households—generate nonmaterial services using material (as well as labor) inputs. Again, these are fairly obvious. They include transportation, wholesale and retail trade, finance, insurance, and education. Some difficulties arise, again, where production or conversion activities are lumped in practice with distribution or other service activities. Electric utilities are the prime example. The ideal solution, of course, would be to separate electric power generation and distribution into separate sectors.

Unfortunately, the generation of services by material goods in the government sector and within households is not covered by any standard economic classification system. However, this topic cannot be ignored if we are attempting to optimize or even, simply, to "map" the entire sequence of transformations from raw materials to final services and consumer welfare. Hence a considerable amount of attention will be directed to this point hereafter.

Taxonomic Considerations[†]

To reach any general conclusions about the future competition between materials, we badly need a means of sorting out what competes with what. In short, we need a classification of uses of materials and of materials by use.

At the most disaggregated level of classification, there are essentially two self-consistent devices. The

[†]In the United States, the Standard Industrial Classification (SIC) is familiar and will be used exclusively. Internationally, the International SIC (ISIC) is a close equivalent.

[†]This section is taken from Reference 38.

first and most rigorous approach is the method of chemistry: to define each material in terms of its exact elemental constituents and its physical form. The alternate device is to use an established commodity–product classification, such as the SITC[39] or the ICGS.[40] Each method has both advantages and disadvantages, which deserve brief mention.

The more precise method starts with a list of the chemical elements and compounds. Such a list can be found in standard reference books. It is naturally very long and can be omitted here since it has little intrinsic interest.

One possible classification of materials is by physical form. There are at least five simple forms (homogeneous solid, granular solid, filamentary solid, liquid, and vapor) and fourteen composite forms. Several of the latter are of special interest:

Laminated solids	plywood
Granule-impregnated solids	concrete
Granule suspension in liquid	slurry
Filament mesh	woven cloth
Liquid droplet suspension in liquid (emulsion)	paint, cream
Liquid droplet in vapor (aerosol)	deodorant
Vapor bubble in liquid (foam)	soapsuds

For a more complete listing of simple and composite forms, see Table 3.5.

Another possible classification is by electrochemical type. The six categories listed are not distinguished as clearly as might be desired, since complex structural characteristics are involved. The selection of water as the *sine qua non* of solubility is based on the ubiquity of H_2O on the surface of the Earth and in all aspects of human endeavor. Other polar solvents such as ammonia (NH_3) might be mentioned, but most compounds that dissolve in water will also dissolve in ammonia. The alternative focus on solubility in nonpolar hydrocarbons is simply based on the fact that hydrocarbons (such as benzene or CCl_4) are the most familiar nonpolar solvents, and the fact that nonpolar compounds tend to dissolve in nonpolar solvents and *not* in polar

Table 3.5 Classification of Materials by Form

Simple Forms	Example
Solid	wood, stone, glass, metal castings
Granular or powder	sand, sawdust, metal powder, cells
Filamentary	hair, feathers, vegetable fiber, muscle, synthetic fiber, wire
Liquid or fluid	water, fats and oils
Vapor or gas	air, steam

Selected Composite Forms	Example
Solid–solid (laminated structure)	plywood
Granule in solid matrix	concrete
Filaments in solid matrix	fiberglass, fiberboard
Filament matrix (mesh)	woven cloth, wire screen
Liquid in solid matrix	wet sponge
Vapor in solid matrix	foam rubber, pumice, coke
Granule in liquid (slurry)	mud, paint, syrup, brine
Granule in vapor	dust cloud
Filament in liquid (gel)	gelatin, silica gel
Filament in vapor	cotton fluff
Liquid droplets in liquid (emulsion)	cream, gravy
Liquid droplets in vapor (aerosol)	aerosol spray (water, paint)
Vapor bubbles in liquid (foam)	soap suds, micro-gas dispersion
Vapor bubbles in solid (solid foam)	styrofoam

Source: Reference 41.

Table 3.6   Electrochemical Classification, Basic Material Forms

Type	Examples					
	Solid	Granular	Filamentary	Mesh	Liquid	Vapor
Metals (conductors)	metals or alloys	metal powder or filings	metal whiskers or wire	wirescreen	molten metal	N.A.
Polar electrolytes (soluble in H_2O)	metal chlorides, nitrates, sulfates, hydroxides, some phosphates, some carbonates	powdered salts			fused salts, e.g., H_2SO_4, HNO_3, H_2CO_3, NH_4OH	HCl, HF
Polar nonelectrolytes (soluble in H_2O)	sugars, amino acids, soap	powdered: sugar, soap, amino acids	spun sugar		alcohols, H_2O, ethers, organic acids, aldehydes	SO_2, SO_3, NO, NO_2, NH_3, O_2 ↔
Nonpolar compounds (soluble in nonpolar HC, e.g., benzene)	resins, plastics, waxes, fats	tetraethyl, lead, iodine	many synthetic polymers (e.g., rayon)	rayon cloth	paraffins, fats, olefins, aromatics, bromine	chlorine
Insoluble, organic (insulators)	rubber, Teflon,© lignin, coal, some proteins, cellulose	powdered coal, etc.	hair and wool, silk, muscle fiber, cotton, jute, wood pulp (cellulose)	wool, silk, cotton, linen cloth, sponge	latex fluorocarbons	Freon© refrigerants
Insoluble, inorganic (insulators)	metal oxides, silicates, phosphates, carbonates, glass, igneous rocks	powdered oxides, powdered glass, sand, clay, soil, gravel, sulfur	asbestos, glass fiber	asbestos or glass cloth	fused rock, molten glass, liquid nitrogen	helium, nitrogen, carbon dioxide, carbon monoxide, hydrogen, fluorine

Source: Reference 41.

solvents. Note that there are organic compounds (and even hydrocarbons) in each group. The "insoluble" materials, on the other hand, seem to divide most conveniently into the organics and inorganics. The former group includes rubber, Teflon,© lignin, coal, cellulose, and many proteins. The latter comprises igneous rocks, gem stones, glass, many metallic oxides, silicates, carbonates, sulfides, and so on. It also includes such exotic materials as semiconductors. Insoluble organic solids are often used as thermal or electric insulators, fabrics, and heat transfer fluids, but seldom as structural materials. Inorganic solids tend to be harder and less flexible, more often used for structural purposes.

Table 3.6 shows a cross-classification of the two. This yields (in principle) a matrix of 6×21 categories, each of which defines a realm of competition. Unfortunately, there is no real guarantee that materials in different categories cannot also compete with each other, so the type of physical–chemical classification already described is aesthetically satisfying but not necessarily always appropriate. An alternative approach for application to modeling the real world is a commodity-based classification system that is sloppier in terms of its fundamental distinctions but that is based on actual patterns of production, exchange, and usage. Here, the starting point might be a list of commodities that are frequently involved in international trade. (The complete list is, again, far too lengthy to include here.) One obvious advantage of the SITC or ICGS is that they are numerical codes that can, in principle, be expanded infinitely by adding more digits as new materials and products are developed. Another advantage of the SITC is that its first four digits coincide with the ISIC.† The third (and last) advantage is that these systems are familiar and widely used, and statistical data based on these classifications is routinely compiled on an international basis.

These practical advantages are formidable, but even in combination, they fail to overcome the fundamental weakness of these systems: their essential arbitrariness and failure to reflect real physical relationships. Neither system, for example, offers any assistance whatsoever in identifying and (taxonomically) aggregating materials that are environmentally dangerous or toxic. (Both freely aggregate

toxic and nontoxic materials in the same category.) Both systems aggregate outputs of totally different processes; even sometimes lumping together inputs and outputs. On the other hand, the SITC sometimes distinguishes commodities that are physically identical and produced by similar means, based on nonphysical distinctions. For example, natural rubber produced on plantations is classified differently from rubber gathered in the forest.

Unfortunately neither of the foregoing approaches is adequate. There are enough problems with both to suggest the possibility of a compromise between them. Like most such compromises, the scheme described next incorporates some of the disadvantages, as well as some of the advantages, of the other two self-consistent approaches. One of its disadvantages is that it is so pragmatic that it is difficult to state the precise principles on which it is based. Suffice it to say that it contains elements of the "method of chemistry" and of the "method of economics," while also introducing some original ad hoc features. Basically, it attempts to classify materials by use, while simultaneously classifying finished materials uses in terms of the materials attributes required.

Finished materials can be divided, first of all, into two major groups, namely *durable* materials destined for use in products designed to be used repeatedly over a period of several years, at least, and *consumable* materials, destined for use in products that are normally "used up" as they are utilized. Most metals, and construction materials, fall into the former group, whereas food products, pharmaceuticals, cosmetics, lubricants, cleaning agents, and numerous others belong in the latter group.

The principal uses of durable and nondurable materials are displayed in Tables 3.7 and 3.8, according to which combinations of properties are mainly involved. Competing materials are listed in italics under each distinct use. Of course, the use categories are somewhat aggregated and many subcategories could be defined for each one, depending on finer differentiation among the required attributes. For instance, rigid pipes could be subdivided according to whether hot or cold water or some other liquid is to be carried, whether the pipe is above or below ground, or whether ease of installation is more important than maximum lifetime. Different materials would sometimes be preferred depending on which of these subcategories was applicable.

---

†The International Standard Industrial Classification system (of industrial sectors) is very similar to the United States SIC. This does not apply to the ICGS.

Table 3.7 Uses of Durable Materials, Based on Properties

	1	2	3	4	5	6	7	8	9	10	11
Major properties, durable materials	Incompress-ability, inertia	Rigidity, weather-resistance	Rigidity, impenetra-bility	Rigidity, strength, toughness, temperature resistance	Flexibility, inelasticity, toughness, impenetra-bility	Flexibility, elasticity, toughness, impenetra-bility	Flexibility, inelasticity, toughness, tensile strength	Flexibility, inelasticity, toughness, absorbency	Thermal/electrical conductivity, Moderate flexibility, temperature resistance	Thermal/electrical non-conductivity (dielectric capacity)	Flexibility, toughness, impenetra-bility, adhesion
Use (category)	ballast/filler/aggregate	load-bearing beams, rods, spars, spokes	structural skin (exterior wall, roof)	engine parts, block, pistons, etc.	wall, floor covering	seals, gasket, bearings	load-bearing cables, wires	materials for printing books	motor generator parts	non structural insulating material	binders
Material (type)	crushed stone, slag, gravel; hydraulic fluids; high-mole. silicones, oil	concrete, steel, wood (for furniture); load-bearing walls; concrete, brick, masonry, wood; pavement; asphalt, concrete, stone (cut), tile (ceramic); +; toughness and strength; frames for vehicles machines bridges; steel, cast iron, reinforced concrete, aluminum alloy	wood, plywood, aluminum, asbestos; shingle plastic; interior wall, floor, ceiling; plaster plasterboard plywood plastic; constant dimension containers: dishes, tubs, tanks; cast iron, lead, copper, galv. iron, PVC, ceramic, concrete; windows; transparency; glass	cast iron, cast aluminum, cast ceramic; axles, bearings, gears, cams, turbine, wheels, valves; steel forging, alum. forging, ceramic, superalloy; burner parts manifold; steel, stainless steel, superalloy, ceramics; +; hardness; cutting tools; carbide alloy, sinter	paper, plastic, linoleum, carpet, felt, woven fabric; furniture covering; plastic, leather; tubes, hoses; plastic (PVC)	rubber; shoe soles, heels, cushions; foam rubber, solid rubber, plastic, leather; tires; rubber	steel; rope, string, cord, net; cotton, jute, hemp, nylon, rayon; power transmission belts; rubberized fabric, chain; +; weather resistance; wire mesh (screen, fence); aluminum, plastic-coated steel	paper plastic "mat"; materials for clothing furnishings etc.; cotton, wool, synthetics, paper, leather	copper, aluminum; electrical wire conductors; copper, aluminum, silver; +; liquidity; secondary battery electrolytes; acid, alkali, molten salt, organic conductors; +; impenetrability; heat exchanger parts; copper, aluminum, steel	feathers, down,; asbestos, fiberglass, plastic foam, PCB, silicones (fluid); +; rigidity; load-bearing insulators; porcelain (ceramic), glass	tar asphalt, plastic (epoxy), cement, sulfur; coatings; plastic, heavy oils; +; nonconductivity; electrical insulation for wire

Source: Reference 41.

58

Table 3.8 Uses of Nondurable Materials

Critical property combination, nondurable materials	Solids			Nonsolids						
	Rigidity, impenetrability	*Flexibility and inelasticity, toughness, impenetrability*	*Flexibility and inelasticity, toughness, absorbency*	*Surface modification*	*Solvent power*	*Thermodynamic properties*	*Chemical reactivity*	*Biological activity— antagonistic*	*Biological activity— supportive, synergistic*	*Fuels and additives*
Use (category)	constant dimension containers, cans, drums, boxes	variable dimension containers bags, tubes, sacks	absorbent tissue, padding	wetting agents, detergents, cleaners	solvents alcohols, hydrocarbons, halogenated HC	antifreeze	bleach; acid; alkali; water softener; primary cell	antibiotics, bactericides, and fungicides; herbicides; rodenticides; insecticides; food preservatives; other pharmaceuticals	fertilizers; vitamins and amino acids; sweeteners and flavors; growth hormones; foods	gas; liquid; solid
Material (type)	*steel tinplate aluminum, plastic foam, wood boxes, paperboard*; *+ transparency*; *bottles*; *glass, plastic*	*paper, aluminum foil, tin (tubes), woven fabric, plastic film*; *+ transparency*; *clear film*; *plastic*	*paper, cotton lint*; *+ toughness*; *materials to print on groundwood (newsprint) paper*	*flocculants*; *+ low viscosity*; *lubricants*	*+ low viscosity*; *spreaders*	*ethylene glycol, propylene glycol*; *snow-melting agents $NaCl$ $Ca(Cl_2)$*; *freezing agents*; *ether aerosol propellants F_{11}, F_{12}, CO_2*	*$CaHClO_3$*; *H_2SO_4 HCl HNO_3 HF*; *$NaOH$, KOH, $Ca(OH)_2$ $NaHCO_3$*; *$+$ conductivity*; *electrolytes*			

Source: Reference 41.

59

On the other hand, the table already subdivides some use categories. Thus pipes and tubes appear in two separate columns, depending on whether *rigidity* or *flexibility* are required. Similarly, furniture coverings are found in separate columns, depending on whether impenetrability (i.e., impermeability to dirt) or *absorbency* (i.e., ability to soak up moisture) are more important. In the first case, a plastic such as PVC or polyurethane might be used, whereas in the second case a woven or knitted fabric might be preferred. Note that leather—an expensive material, unfortunately—is found in both columns because it combines both properties fairly well.

Table 3.7 does not include all uses of durable materials. It specifically omits most of the uses that depend on thermodynamic, electromagnetic, optical, and nuclear properties. These are discussed separately, because of their special importance in current "high technology" applications. One might mention the following:

- *Pigments.* These materials selectively absorb and reflect specified optical wavelengths. Natural pigments include titanium dioxide, zinc and lead-oxides, cadmium, cobalt, chromium, graphite, aluminum, gold, and other metallic pigments, plus synthetic organic dyes such as the anilines. The total market amounts to several billions of dollars per year.
- *Optical materials.* This classification includes photon absorbers and emitters and photoreactive substances in optical and near-infrared wavelengths. These are of special interest with respect to illumination, photography, xerography, surveillance, TV, optical communications (including optical fibers, lasers, scanners, LEDs, and LCDs), photosynthesis—both natural and artificial—and the development of solar cells to convert solar energy into electricity. Silver is the biggest single material input to optical technology and is used for photographic film, but other materials include tungsten, used as filaments in incandescent lights, as well as in fluorescent coatings for fluorescent lights and TV tubes, neon, sodium and mercury vapor; gallium (e.g., in gallium arsenide, and gallium phosphide diodes); various rare-earth phosphors; aluminum, the key to the xerography process; ultrapure silicon (for solar cells), synthetic rubies for lasers; and a host of other exotic materials. Altogether, production of these materials also amounts to several billion dollars annually and is growing fast.

- *Nuclear materials.* In this category are included all potentially fissionable isotopes, such as, plutonium, U-235, U-233, as well as naturally occurring isotopes that can be converted into fissionable form by neutron absorbtion, namely, U-238 and T-238. Also, we include fusion isotopes deuterium (H_2), tritium (H_3), Li-6, and various other radioisotopes with commercial or scientific uses, such as radium and Co-60. Also, for convenience, we include other materials widely used by the nuclear industry such as lead and zirconium. The manufacture and processing of nuclear materials for nonmilitary use is, or soon will be, a multi-billion dollar business.
- *Magnetic materials, ferrites, and superconductors.* In this category are included all materials used for permanent magnets (such as AlNiCo and the newer cobalt–cerium alloys); materials used for ferrite cores for high-speed computer memories, magnetic drums, disks, and tapes for computers, as well as magnetic audio and video tapes (iron and chromium oxides); and materials used for superconducting magnets (alloys of niobium, zirconium, and titanium). Future uses of superconductors for other purposes, such as electric power transmission, would also be included. Total output at present is several hundred million dollars per year.
- *Semiconductors.* In this category are included all materials (other than photo-emitters, absorbers, and photoreactors) used in solid-state electronics. Germanium and silicon are still the most important semiconductor elements, though a large number of other elements find significant uses. Total output at present is several billion dollars annually.
- *Electrodes.* In this category are included anodes and cathodes for both primary and secondary batteries and electrolytic cells. Materials used today are lead, zinc, graphite, nickel, and manganese. Future materials may include lithium, sulfur, magnesium, and others. Total output of materials for these purposes is several billion dollars per year.
- *Refrigerants.* These are chemically inert, low-molecular weight, fluorinated compounds (usually F_{11} and F_{12}), popularly known by the Dupont trade name "Freon," though other manufacturers sell these compounds under different trade names. These compounds are useful because their thermodynamic characteristics—vapor pressure-temperature curve, heat capacity, heat of

vaporization—are suitable for operating a compression cooling cycle within roughly 15°C above and 30°C below average ambient temperatures (20–25°C). They are widely used in household, commercial, and automotive air conditioners, refrigerators, and freezers. Current production for these purposes is several hundred million dollars annually. A roughly equivalent annual production goes into propellants for aerosol sprays (deodorants, hair sprays, and insectides). As this is written, the latter application is under threat of ban by EPA on the grounds that fluorocarbon molecules may constitute a serious danger to the ozone layer of the mesosphere.

Miscellaneous. In addition, there are numerous highly specialized materials used in rather small quantities, such as X-ray emitters and absorbers, ultraviolet emitters and absorbers, dental alloys, catalysts, ultra-high temperature ablative materials for nose cones and rocket engine liners, filter materials, absorbers, ion-exchange membranes, high temperature electrical connectors, and welding rods. The noble metals, especially platinum, palladium, nickel, mercury, graphite, and tungsten are important.

Another important aspect of the uses of materials is the extent to which they are subsequently recoverable. Recoverability is fundamentally determined by the amount of free energy required to separate the material from its waste form into a usable form. Two kinds of recycling must immediately be distinguished, namely,

1. Recycling of waste material as an ore requiring beneficiation, reduction, refining, and fabrication.
2. Rebuilding or modification of discarded objects to be used as such.

For instance, recovery of metal cans, paper, or glass from mixed municipal refuse is an example of category 1. However, recycling of worn out lead-acid storage batteries, tires, engines, or transmissions would be in category 2. The energy (and negentropy) required for recycling of the second kind is substantially less than for the first kind, but much more information about the waste is required. To recycle tires, strict standards of tread wear must be adhered to. To rebuild engines or transmissions, *exact* mechanical details of the original design are required.[†]

[†]This means model number, catalog numbers for parts, and similar identifying numbers.

As might be expected, some uses of materials are such as to invite nearly complete recycling (steel rails, telephone cables, and automobile storage batteries are in this category). Others are compatible with partial recycling (e.g., tires and newspapers), whereas some uses are such that recycling is intrinsically impossible (e.g., fuel additives and pigments for paint). Table 3.9 is a rough allocation of finished materials into recoverability classes. Table 3.10 identifies some natural competitors and inter-material substitution possibilities.

Aggregate Classification of Elements, Based on Environmental Hazard

Unfortunately neither chemical–physical nor trade classifications nor classifications based on functional use offer much guidance with respect to environmental risk. Yet this topic is of vital interest. The question we address here is, simply, how should the chemical elements be aggregated in such a way as to group together those elements (and compounds derived therefrom) with key environmental attributes in common? On the one hand, certain elements are widely available on the Earth's surface, easy to extract, and relatively compatible with living systems. Other elements are important constituents of environmental processes but pose special problems for human utilization. Still others are essentially not compatible with natural processes. The latter elements are also, in general, not environmentally recycled and not widely available in concentrated deposits. They are, in consequence, both intrinsically scarce and risky to use (at least, dissipatively). Thus conservationist and environmentalist motivations coincide in urging reduced overall extraction of these materials—supplemented by more intensive efforts to recycle them—and introduction of strong disincentives to utilize them in applications that will result in significant losses back to the environment.

Group I Elements:

hydrogen	sodium	silicon	iron
carbon	potassium	magnesium	
oxygen	calcium	aluminum	

Group II Elements

| nitrogen | sulfur | phosphorus | chlorine |

Group III consists of all other elements.

Table 3.9 Materials Classification by Recoverability

Material	Nonrecoverable	Partially recoverable	Fully recoverable/reusable < 5 years	Fully recoverable/reusable > 5 years
Wood	boxes, crates, lathe, fence frames	furniture, panels, heavy beams siding		railroad ties
Paper	wrapping paper, household paper	telephone books, books, newsprint, magazines, computer paper		
Cotton	lint, gauze, string	shirts, sheets		
Wool	most clothing	blankets		
Synthetic, mixed fibers	all			
Foam rubber	all			
Solid rubber	belts, tubes, erasers, heels, etc.	automobile tires	truck and aircraft tires	
Foam plastic	all			
Plastic film	all			
Plastic moldings	most	some items (e.g., trays)		
Plastic pipe		?		
Plastic sheet extrusions		?		
Glass	window glass	bottles, auto glass		?
Iron steel	chemicals	galvanized sheet, cans, household, appliances, wire products, hardware	drums	pipe (water, gas) tanks, rails, girders, machines, car/truck bodies
Copper	chemicals (e.g., copper sulfate)	wiring from machines, shell/cases, brass hardware		telephone/electric cable, water pipes, auto radiators
Zinc	primary cells, galvanizing chemicals	castings		
Lead	bullets, solder, TEL, pigment for paint		auto/truck batteries	pipes, acoustic sheet, shielding (bricks)
Aluminum	foil, plating	cans, frames, hardware		siding and roofing, aircraft bodies
Nickel, chromium	chemicals, plating pigments, plating	batteries, stainless steel appliances, cutlery, kitchenware		stainless steel tanks
Cobalt, manganese, vanadium, molybdenum niobium	pigments, chemicals, electrochemical cells	superalloys, high-temperature alloys, high-speed alloys		
Tungsten	welding rods, phosphors filaments, tire studs	high-speed alloys, carbide bits		
Cadmium	pigments, plating, chemicals	battery anodes		
Tin	plating, tubes			
Mercury	dental, pharmaceutic, electrical	catalysts		
Silver	photo film, dental alloys	silverplate		sterling, cutlery jewelry, coins
Gold	dental castings, electrical connectors			jewelry
Platinum		catalysts		

Source: Reference 41.

Table 3.10 Substitution Possibilities, Major Metals

Competing metals	Functional use	Likely substitute(s)	Investment time required	Environment/energy impact
Aluminum	External shell for static structures (containers, siding, roofing, etc.)	painted wood or fiberboard, corrosion-proofed or plated iron sheet, asbestos tile/sheet, glass (containers or windows), vinyl plastic	all are currently competing; rapid substitution	Depends on process detail. all alternatives require less energy to produce, but are difficult to recycle.
Aluminum alloy Carbon Steel	Support/shell for vehicles (automobiles, aircraft)	each other, magnesium alloy, fiberglass/ABS plastic	steel is currently dominant; for plastic must replace metal stamping facilities; for aluminum or magnesium must increase capacity	steel vehicles are heavier and use more fuel (than aluminum) magnesium or plastic vehicles would be lighter and use less fuel, but less durable
Aluminum Copper	electrical and thermal conductor in wires, heat exchangers	each other	currently competing; rapid substitution	copper is a more efficient conductor, lower energy losses.
Aluminum Brass Zinc	small castings, with low mechanical strength and good corrosion resistance	each other, plastic	currently competing; rapid substitution	zinc and brass use less energy but come from sulfide ores, plastics would not be recyclable
Manganese	as anti-brittleness alloying element	nickel, chromium, molybdenum nitrogen	all are currently used in different combinations, depending on specific application.	depends on details of application,
Molybdenum Tungsten	as alloying elements for steel to provide hardness at high temperatures (e.g., for cutting tools)	sintered tungsten carbides, or other (silicon, titanium) carbides; corundum (alumina); other means of metal removal (e.g., electric spark discharges)		Superalloys or ceramics permitting higher temperature operation, e.g. for turbines will result in substantial efficiency savings: ditto for cutting tools. alloy steel is not easy to recycle.
Vanadium	as an alloying element for structural steel	manganese, titanium, molybdenum, columbium	rapid substitution occurs where close equivalents exist, but substitution is slow where alternatives involve substantial adaptation (e.g., of machine tools, design specs, etc.)	
Chromium Cobalt Columbium Nickel	as alloying elements to provide toughness, corrosion resistance (stainless steel), and high-temperature strength for steel and "superalloys"	others in the group; aluminum, titanium, silicon ceramics, (e.g., alumina)		
Chromium Nickel Zinc Aluminum	non-corrosive or decorative plating elements for carbon steel	each other; tin, cadmium, titanium, polymer coatings, stainless steel, aluminum metal	metals and stainless steel are currently competing; rapid substitution between metallic plating elements.	chromium and cadmium plating processes lead to severe water pollution problems; zinc and cadmium are extracted from sulfide ores; aluminum and titanium cladding involve more energy; polymer coating would be best for recycling.
Lead	for automotive SLI battery	nickel–iron, nickel–zinc, nickel–cadmium	no exact equivalent exists, hence change would be slow	not much since batteries are almost 100% recycled

Source: Reference 41

The key characteristics of Group I as a whole are (a) wide dispersion and essentially unlimited availability on the Earth's surface (except, possibly, potassium) at or near the present extraction costs and (b) that minimal environmental problems are associated with their use. The first three of these materials (H, C, O) are recycled naturally by biological processes. They are, obviously, the major structural elements in living systems. Dissipative and dispersive uses of these materials—except, possibly, fossil hydrocarbons—are neither environmentally dangerous nor unsound from a conservationist point of view. The remaining seven elements are by far the dominant components of inorganic structural materials used on Earth today—both metals and nonmetals—from a weight, mass, or value standpoint.

The pollution problems associated with Group I are silting and erosion, soot (unburned carbon particulates), CO, vaporized hydrocarbons, BOD,[†] and oil spills. The first is due to careless and easily controlled agricultural and construction practices; the next three are due almost exclusively to inefficient combustion of fossil fuels. BOD arises from waterborne disposal of untreated or partially treated municipal sewage, agricultural food processing, or pulp and paper manufacturing wastes. The last is partly due to shipping accidents and partly to tanker operating practices. All of these problems are relatively easy to reduce by a combination of straightforward technology and regulation.[‡] Evidently, all of the elements and compounds in Group I can be assimilated and recycled by natural environmental processes. Problems arise essentially from quantitative excess and embalances, not from fundamental incompatibilities.

Group II elements are also widespread and readily available on the Earth's surface, except for phosphorus. Nitrogen is recycled in the environment by natural biological processes. They are uniquely troublesome in that they are all "critical" (though quantitatively minor) components of living systems. Phosphorus and nitrogen are often added to soil as fertilizer to increase agricultural productivity. Many compounds of Group II elements with carbon, hydrogen, and oxygen do not occur in nature, yet are extremely biologically active—sometimes highly toxic or carcinogenetic.

[†]BOD = Biological Oxygen Demand
[‡]The economic costs and/or the political difficulties may be significant in some cases.

Thus nitrogen is a critical component of all amino acids—the building blocks for protein required for all living cells. Yet closely related compounds, the nitrosamines, have been identified as among the most carcinogenetic of all chemicals. Many nitrogen-containing components are highly toxic to animals, beginning with a simple hydrogen cyanide (HCN) and many derivatives, as well as gaseous oxides of nitrogen NO_x, nitrous and nitric acids, and derivative compounds, especially nitrites. Sodium nitrite, used to retard spoilage in bacon, is both toxic and probably carcinogenic. Reaction between oxides of nitrogen, hydrocarbons, and ozone in the atmosphere also produce highly toxic and possibly carcinogenetic pollutants like peroxy acetyl nitrate (PAN).

The sulfur story is similar. Pure sulfur is toxic: It was formerly used widely as a fungicide (as was copper sulfide—CuS). Hydrogen sulfide (HS) is highly toxic, along with the derivative mercaptans (CH_3, HS). Sulfur oxides (SO_x) are toxic both in gaseous and acid (H_2SO_4) form as are carbon disulfide (CS_2), and many metallic sulfides and sulfates such as copper sulfate. Some sulfur-containing organics—sulfonamides, for instance—have valuable medicinal uses precisely because of their toxicity to living micro-organisms. Sulfur is not scarce. Notwithstanding some recent talk of scarcity of natural elemental sulfur, the world faces a vast sulfur oversupply in the foreseeable future because of the necessity of removing unwanted sulfur from petroleum and coal before burning.

Phosphorus is sometimes described as the "bottleneck of life," since it is relatively scarce on the Earth's surface, yet absolutely required for many fundamental biological processes. It is available mainly from deposits of ore where it is combined with calcium, silicon, oxygen and fluorine (in trace amounts). Phosphorus is a key component of DNA and RNA, the self-reproductive material in each living cell. A phosphorus compound, adenosine triphosphate (ATP) is critical to the process by which animal muscles release stored energy, and by which nerves send and receive signals. Many enzymes as well as amino acids and proteins contain phosphorus. Nor surprisingly, perhaps, phosphorus-containing organics such as phosphoryl cholinesters can also be highly disruptive to organized life processes. Many of the most highly toxic insecticides in use today (such as sevin, malathion, and parathion) are phosphorylated organics. Fortunately, these materials break down fairly quickly and do not linger in the environment.

Chlorine has a somewhat lesser role in living systems, though either sodium chloride (NaCl) or potassium chloride (KCl) are an absolute requirement for human nutrition. But pure chlorine is highly toxic, which accounts for its use in water treatment, as is hydrogen chloride HCl and other gaseous chlorine compounds such as "phosgene" (CCl_2) as well as a number of commonly used solvents such as carbon tetrachloride (CCl_4). Many chlorinated hydrocarbons are highly toxic, including the well-known insecticides DDT, endrin, dieldrin, kepone, and the herbicide 2-4-D. Vinyl chloride, moreover, used to make the plastic PVC, has recently been discovered to be very carcinogenic. The polychlorinated biphenyls (PCBs) have also recently been implicated as dangerous to the natural marine environment. The key reason seems to be that some toxic chlorinated hydrocarbons are preferentially accumulated in living systems until dangerous levels are reached; on the other hand, they are broken down very slowly (by bacterial action) in the natural environment.

Clearly, Group II elements are both highly important and valuable, and also highly dangerous in many forms. This suggests that special care should be exercised in their manufacture and use. It also suggests that extra effort should be devoted to monitoring their stocks and flows.

Group III elements could be further divided into those, mainly metals, that are economically significant and those that (for the present) can probably be ignored. Many of the economically significant metals or semimetals are geochemically classed as chalcophiles (sulfur-loving) and are often found in nature as sulfide ores. This group includes copper, lead, zinc, nickel, mercury, and silver plus several by-product elements like antimony, cadmium, arsenic, thallium, and bismuth. Virtually all of these elements have toxic compounds, though some—like mercury and cadmium—are much more toxic than others. Zinc, copper, and silver are among the least dangerous. The so-called lithophile metals, which occur primarily as oxides, are generally less toxic.[†]

[†]Chromium is exceedingly toxic in one of its ionic forms, and beryllium dust is highly toxic if inhaled into the lungs. Also, it goes without saying that most of the radioactive elements are at least somewhat dangerous. Plutonium, in particular, is highly toxic and carcinogenetic, as well as fissionable.

The major nonmetals in Group III are the halogens: fluorine, bromine, and iodine and the inert gases (helium, neon, and argon). Of these, fluorine is both the most significant and the most hazardous, as well as being intrinsically scarce. Semimetals such as germanium and selenium are also becoming economically important and may pose some unexpected environmental problems in the future.

Classification of Transformations

As before, there is a choice of several levels of aggregation and a choice between the method of chemistry (or physics) and the method of economics.

The method of chemistry views a transformation as a change in one of the state variables of a system. Relevant state variables might include the following (among others):

- Location.
- Velocity or momentum.
- Temperature, heat content, and free Energy.
- Pressure, torsion, tension.
- Phase.
- Form.

A possible list of elementary transformations (or microprocesses) is attached as Appendix 3-B. The list is complete if and only if any *macro-process* is decomposable unambiguously into sequences of elementary transformations. However, even a rigorous demonstration of decomposability would be of little practical significance, unless we could also demonstrate the inverse procedure, namely, that viable macro-processes can be "synthesized" from more elementary components, by application of some optimization algorithm subject to a set of constraints.

Ideally, the synthesis procedure would follow a scheme such as is shown in Figures 3-B.1 through 3-B.6. When the problem is stated in these terms, however, its true difficulty and complexity can be grasped more easily. Problems chiefly arise from the fact that the constraints governing elementary transformations or reactions between material states are *not* independent of the initial or final state of the material. (Indeed, these rules are also often functions of physical or chemical composition as indicated by Tables 3-B.1 and 3-B.2.) In ordinary language, this is equivalent to the unsurprising observation that the limiting factors governing—for example —the freezing of water into ice cubes are different from the limiting factors governing the "freezing" of molten iron into castings. The similarity, in principle, between these processes is obvious, but so are the differences in practice. These differences are, of course, critical to process engineering and process economics. It is regrettable but clear that the state-of-the-art of process synthesis today, is not advanced enough for large-scale application.[†]

Thus we are led to the more aggregated method of economics as the only feasible alternative. This approach attempts no a priori definition of a process in terms of fundamental principles. Rather, a macroprocess (to use the former designation) is defined as a technologically based activity that converts purchased material inputs to salable material outputs, on a scale corresponding to a *firm* or *establishment*.[‡] (Production of pure services such as transportation is omitted from further consideration for conceptual reasons). The only "process" involved, from a materials/energy point of view, is *consumption* of fixed capital and of labor-saving fuels and materials. To facilitate congruence with the terminology of process engineering, the term "*economic unit process*," which underlines its comparability to conventional engineering usage, is introduced.[§]

While it suffers to some extent from apparent arbitrariness, it should be recognized that the economic approach to classification—whether of commodities or of processes—is not unreasonable from a physical point of view. Indeed, it is a significant empirical fact that for any given basic or finished material, at any given time, there are typically only two or three alternative economic processes—seldom more. Two other pertinent facts are worth noting:

1. The viable processes in any industry are well known in the industry. In practice, there is not much difficulty of identification or characterization.
2. There are likely to be a number of competing *potential* future processes (at least, in a growing industry). The competition occurs at the level of

---

[†]The challenge of developing heuristic principles for synthesizing processes from elementary transformations has not been totally overlooked, however. For instance, an approach utilizing an interactive computer program (AIDES) to assist a chemical process-designer has been developed by Rudd and collaborators.[43]

[‡]The term "establishment" is used here as in the Census of Manufactures.

[§]The terms "unit process" and "unit operation" are common in chemical and sanitary engineering; the former connotes a chemical or biological transformation, whereas the latter connotes a physical transformation. Both terms imply a characteristic configuration of equipment.

paper studies, laboratory experiments, and pilot plants. Not until a process has proved successful at the pilot stage can it be considered viable. Evidently, the successive development stages involve testing a large number of alternative configurations, pathways, design principles, and materials. Only the survivor(s) of this selection process reach the production stage.

The design of a viable industrial process, therefore, represents a suboptimal choice from among an infinite range of initial or a priori possibilities. (Comparable statements can be made, on the one hand, for viable *products*—or services—and on the other hand, for viable biological *species*.) Of course, further development and optimization occurs, even after the process is adopted, as numerous minor incremental improvements are made in design, operating conditions, control, and so forth. This postdesign phase of gradual final optimization is commonly referred to as the "experience curve."

The set of all possible transformation processes is "dense" in the mathematical sense, whereas the set of near-optimal regions—given an objective function and a set of uncertain technological constraints—is "discrete." The distribution of near-optimal regions—that is, economic processes—is, moreover, a reflection of the fundamental technological possibilities as governed by the properties of materials and the laws of chemistry and physics. Thus the arbitrariness of the economic approach to classification is apparent, not real. It enables us to focus attention on a finite set of possibilities, rather than a continuum.

3.4 CONCLUDING REMARKS: THE NEED FOR A NEW PARADIGM

From the point of view of materials and energy, the economy is a system of transformations that convert raw materials and natural resources (both renewable and nonrenewable) into "final" goods and services. This transformation is unidirectional: Crude materials and energy are extracted, refined, processes, recombined in new forms, fabricated, and assembled into articles of commerce of "infrastructure." These products are not "consumed" in a physical sense. Rather, they render specific services to "consumers," after which they are discarded as wastes.

From the traditional point of view,[44, 45] the economy is a system for generating and allocating final services that maximize their utility to consumers (consumer welfare). The utility-maximizing combination and allocation of services is, of course, the Pareto optimum. The central problem of economics is to determine the most efficient allocation of resources to achieve this optimum.

The word "resources" in the previous sentence implies factors of production, meaning labor (by skill), capital investment, raw or processed materials, and organized productive entities—firms—or activities.

It has been stated that welfare is attributable to services. Final services are rendered to consumers via six distinct pathways, as follows:

1. Via personal interaction (e.g., with teachers, doctors, and professional advisers).
2. Via *autonomous* material goods (e.g., food, clothing, books, jewelry, hand tools, toys, and furniture).
3. Via *nonautonomous* (interlinked) material goods (e.g., household appliances, automobiles, and housing) and associated network systems (electric and gas utilities, telephone, roads, water, and sewers).
4. From impersonal governmental agencies (e.g., health, safety, and police).
5. From nongovernmental social institutions and civilized society (e.g., religion, culture, language, neighborhoods, and family).
6. From the natural environment (e.g., air, landscape and natural beauty, biological production, and waste recycling and stabilization).

"Autonomous" goods are complete in and of themselves and require no continuing connection with an energy supply, a communications network, a water or sewer system, or any other utility support. Autonomous goods would be (more or less) as useful in an uninhabited or primitive place as in a cosmopolitan city. This is not true of nonautonomous goods.

The first three pathways for delivering final services are potentially subject to marketplace allocative mechanisms, although obviously nonmarket mechanisms can be—and often are—employed. The fourth route is inherently subject to political means of allocation. Some personal services such as education or electric power normally supplied by government *can* be supplied by private enterprises. This is not true of impersonal services, however. Nongovernment social and environmental services (cate-

gories 5 and 6) are allocated in ways that are neither strictly market determined nor political, but a combination of these plus other factors.

Economists are concerned with two issues (at least). First, it is important to evaluate the extent of synergism or (more likely) antagonism between different types of services, especially where they are allocated by different mechanisms. The magnitude and origin of "externalities" on the quality or availability of social and/or environmental services due to materials and energy-intensive activities associated with services of types 2 and 3 must be evaluated, for instance. Second, the theory of welfare economics must clearly recognize the possibility of substitution not only between different services *within* the market-allocated group (1–3) but also between market and nonmarket allocated categories. It is important to understand more fully the extent to which utility derived from one class of services may be a satisfactory substitute for utility derived from another.

To recapitulate briefly, economics has traditionally concerned itself with the optimization of a two stage transformation (or mapping): from *factors of production* to *goods and services* and thence to *utility or welfare*. The first stage of this sequence is optimized in principle by suitable organization of production and allocation of resources (i.e., factors of production). The second stage is optimized by allocation of incomes among consumers and expenditures among goods and services. In this *classical* paradigm, physical characteristics of materials, energy, machines, and processes have essentially been abstracted into mathematical properties of production functions. Indeed, the level of abstraction is so great that the assumed mathematical characteristics are often inconsistent with basic physical laws, as discussed previously in this chapter. Many of the simple resource/environment models fail to satisfy even the First Law of thermodynamics, still less the deeper laws of physics.

It now seems appropriate to introduce a somewhat more elaborate paradigm: a more complex (but also more realistic) sequence of mappings from *exhaustible or renewable natural resources* to *finished materials and forms of energy*; then to material *products*, and *structures*,[†] then to (abstract) *services*, and finally to *utility*. It is noteworthy that the first two transformation stages are essentially physical in nature, whereas

---

†This includes capital goods.

the latter two are not. Of course, the physical products are not consumed in any literal sense: They are recycled as alternative sources of material or they are returned to the environment.

At each stage of transformation, there are a number of combinations of inputs, each of which is *sufficient*, but not *necessary*, to generate one of the outputs.[†] In the first stage—from raw materials to finished materials—inputs also include labor, labor-saving machines (i.e., capital goods), and labor-saving intermediate materials such as fuels, electricity, fertilizers, pesticides, and flotation agents. Each family of sufficient combinations of inputs, taking into account all possible substitutions between labor, capital goods and labor-saving materials, constitutes a *process technology* (or, simply a *process*).

In the second stage—from finished materials and forms of energy to material products and structures —inputs include outputs of the first stage, plus labor, capital, and intermediate materials, as before. Each family of sufficient combinations of inputs, again taking account of all possible substitutions between labor, capital, and intermediate materials, again constitutes a technology. In both physical transformation stages, the concern of economics is to select the optimal "mix" of technologies, given the availability of raw materials (for Stage I) or the output of Stage I (for Stage II), and the relative prices of materials, energy, labor, and capital.

In the third and fourth stages, material goods and structures are translated successively into services and into utility or welfare. Inputs in Stage III obviously include labor as well as the material products generated by Stage II—including, but not limited to, capital goods and consumer durables. In this stage, all material inputs are converted to wastes; the primary outputs are nonmaterial services. Again, a family of inputs *sufficient* to generate a given final service—taking into account all possibilities of substitution between labor and labor-saving machines or materials, as before—can be regarded as a technology.

The fourth stage of the revised paradigm is similar, but not quite identical to the previous one. Maximization of utility is accomplished, in principle, by suitable allocation of incomes among consumers and of expenditures among pure final services. Un-

fortunately, there is no market for pure services as such. Actually, expenditures go for real goods and services, which offer highly selective combinations of attributes in joint-product form. In principle, given enough information on price and income elasticities, and cross-elasticities of demand for such real goods and services, it would be possible to infer the correct shadow prices and elasticities of the underlying pure services or attributes.[†]

Although many professional economists may regard the preceding paragraphs as elementary and, perhaps, a trifle redundant, it is curious that so many economists have assumed that the output of the economy (GNP) is closely related to (and constrained by) the availability of material and energy inputs. Resource economists, in particular, have abetted the public misapprehension in this regard. Yet a major implication of what I have been saying in the last few pages, is this: While materials and energy are essential to the production of goods and most services, *there is no unique irreducible set of material/energy inputs that is necessary to yield a given level of utility, or even a given mix of pure final services.* Each final service can be generated by several alternative technologies (families of material products, structures, labor, etc.). Each material product, in turn, can be produced in a number of ways, using different finished materials, designs, combinations of components, and other similar variables. And each finished material can be derived from a variety of sources, often by several different processes.

This statement does not mean, of course, that energy or materials availability could not be economically limiting, at any given time, in the sense of restricting the viable process options. All real processes clearly have irreducible material/energy requirements per unit of *material* output.

How does it happen that so many economists have seemingly accepted—contrary to their own doctrines—that the GNP is really "limited" in any real sense by energy consumption? For the most part, this myopia probably derives from over emphasis on econometric (i.e., statistical) techniques of analysis and confusion of correlation with causation.[46] Too many economists have failed to admit

---

[†]Of course, the existence of joint products complicates the discussion, but most readers who are aware of the problem can fill in the necessary additional caveats and elaborations.

[†]The relevant theory lies somewhat beyond the scope of this section, and the matter cannot be dealt with adequately here. See the discussion in Section 1.3. Because of the intrinsic difficulties of measurement, this topic is among the more neglected areas of economics.

that long-run elasticities bear no necessary relation to short-run elasticities. But there is also a deeper reason. This is, perhaps, one more instance of the attraction of simplistic axioms. In the eighteenth century, economists believed that *land* was the true basic resource and source of all wealth. Marx and his followers in the nineteenth century rejected this notion and put forward *labor* as the limiting factor of production. Both of these notions have proved to be inadequate to explain the real world. Inasmuch as energy is used to operate machines, which are a kind of substitute for labor, is it not reasonable for the twentieth century to put forward *energy* as the ultimate limiting factor? Certainly this notion has had a certain vogue in recent years.[†] If it were valid, then it should be possible to define an aggregate production function with only one independent variable, energy. And the prices of all commodities should be exactly proportional to their "net energy" content[53] (where net energy would have to refer to "available work" not enthalpy). It need hardly be pointed out that these conditions do not hold true, at least today. It seems highly implausible to most economists that they ever will.

Granting that energy (and materials) are not limiting, ultimate factors of production, though they are co-factors (and constraints) at any given time, leads to two groups of critical and well-defined questions. To begin with, it can be assumed that (for a given population and state of welfare) a fixed set of technological possibilities, a given set of existing capital investments, and a given set of raw material prices will determine an optimum set of technological choices and corresponding raw material inputs. As the price of fossil energy (for instance) is raised eventually higher and higher, *ceteris paribus*, a succession of revised static optimum solutions will be unveiled. This series must eventually converge to the solution that absolutely minimizes energy consumption. Given a certain rate of economic growth (allocated to sectors) and a certain rate of capital depreciation, one can also determine, in principle, the time path over which a "new" static optimum solution will be reached from an "old" one. The pertinent questions for policy-makers are: What op-

tions are available for conserving energy? How much energy would be saved? How much would it cost in public funds? How soon would the benefits accrue? What sectors of society will be adversely affected and how much?

The second critical question pertains to determinants of the path of technological change, under various alternative stimuli. That technology is evolving, and may continue to do so (if society continues to invest in new knowledge) can be taken for granted, notwithstanding the existence of material/energy constraints (e.g., to "negentropy" stocks), as discussed earlier. But both the direction and the rate of change, as they are affected by other political and economic factors, are vitally important. Reverting to the supposition of the last paragraph, How will the rate and direction of technological progress tend to respond, *ceteris paribus*, as the price of energy continues to rise? At what rates can new technologies be developed, in principle, to increase energy supply capabilities or to decrease energy consumption, or both? How will these rates be affected by energy costs, labor costs, interest rates, and the like? What corresponding investment is needed in technology per se? And, what—if any—are the physical limits of technological achievement? Specifically, is there an irreducible ultimate lower limit to the amount of energy that will be needed to yield a given level of utility (i.e., welfare) to "consumers"?[†] Putting the question another way, Is it necessary to postulate a long-run upper limit on welfare or standard of living (per capita or not) as a consequence of the limited long-run renewable energy supply of the Earth?

But, my tentative answer, several paragraphs back, has been no. These questions cannot be answered finally in a book such as this, if indeed there are answers. I call attention to the questions primarily because of their relevance to the design of models and methods of analysis, with which the remainder of the book is concerned.

[†]Among economists who have been cited as fathers of the idea—however inadvertent—one might mention Georgescu-Roegen.[47] The list of noneconomists who have put forward this view is longer, and includes geologists like Hubbert[48] and advocates of "net energy analysis," such as Odum,[2] Berry and Fels,[49] Bullard and Herendeen,[50] Hannon,[51] and Watt.[52]

REFERENCES

1. N. Georgescu-Roegen, *The Entropy Law and the Economic Process*, Harvard University Press, Cambridge, MA, 1971.
2. H. T. Odum, *Environment, Power and Society*, Wiley-Interscience, New York, 1971.

[†]This question was first in a footnote in Section 3.2

3. R. U. Ayres and A. V. Kneese, "Production, Consumption and Externalities," *Amer. Econ. Rev.*, **59** (3) (June 1969).

4. A. V. Kneese, R. U. Ayres, and R. d'Arge, *Economics and the Environment*, Johns Hopkins University Press, Baltimore, 1970.

5. K-G. Mäler, *Environmental Economics—A Theoretical Inquiry*, Johns Hopkins University Press, Baltimore, 1974.

6. T. Scitovsky, "Two Concepts of External Diseconomies," *J. Polit. Econ.*, **62** (April 1954).

7. R. H. Coase, "The Problem of Social Cost," *J. Law Econ.*, **3** (October 1960).

8. J. W. Buchanan and G. Tullock, "Externality," *Economica*, **29** (November 1962).

9. O. A. Davis and A. Whinston, "Externalities, Welfare and the Theory of Game," *J. Polit. Econ.*, **70** (June 1962).

10. R. Turvey, "On Divergences Between Social Cost and Private Cost," *Economica*, **30** (November 1963).

11. W. J. Baumol, *Welfare Economics and the Theory of the State*, Harvard University Press, Cambridge, MA, 1967.

12. S. Carlson, *Pure Theory of Production*, Kelley and Millman, New York, 1956.

13. S. Däno, *Industrial Production Models*, Springer-Verlag, New York, 1966.

14. S. E. Jacobsen, "Production Correspondences," RAND Paper, p. 3974 (November 1968).

15. R. W. Shephard, *Theory of Cost and Production Functions*, Princeton University Press, Princeton, NJ, 1970 (and earlier work cited).

16. V. L. Smith, *Investment and Production*, Harvard University Press, Cambridge, MA, 1961.

17. M. D. Intriligator, *Mathematical Optimization and Economic Theory*, Prentice-Hall, Englewood Cliffs, NJ, 1971.

18. E. G. Korach (ed.), *Technology of Efficient Energy Utilization*, Report of a NATO Science Committee Conference, October 8–12, 1973, Pergamon, New York, 1975.

19. E. P. Gyftopoulos, L. J. Lazaridis, and T. F. Widmer, *Potential Fuel Effectiveness in Industry*, a Report to the Energy Policy Project, Ford Foundation, Ballinger Press, Cambridge, MA, 1974.

20. M. Ross, R. Socolow et al., *Effective Use of Energy: A Physics Perspective*, Report of the Summer Study on Technical Aspects of Efficient Energy Utilization, American Physical Society, January 1975.

21. R. U. Ayres and M. Narkus-Kramer, "An Assessment of Methodologies for Estimating National Energy Efficiency," Winter Meeting of the ASME, November 1976.

22. V. E. McKelvey, "Relation of Reserves of the Elements to Their Crustal Abundance," *Amer. J. Sci.*, **258A** (1960).

23. D. B. Brooks, "The Lead–Zinc Anomaly," *Trans. Soc. Mining Eng.* (June 1967).

24. E. D. Goldberg, "Marine Pollution: Action and Reaction Times," *Oceanus*, **10** (1974).

25. N. Wiener, *Cybernetics*, Wiley, New York, 1948.

26. C. Shannon and W. Weaver, *The Mathematical Theory of Communication*, University of Illinois Press, Urbana, IL, 1949.

27. E. Schrödinger, *What is Life?*, Cambridge University Press, Cambridge, U. K., 1945.

28. D. L. Meadows et al., *Limits to Growth*, Universe Books, New York, 1972.

29. D. L. and D. H. Meadows, *Towards Global Equilibrium*, Wright-Allen Press, Cambridge, MA, 1973.

30. S. Cole et al., *Models of Doom: A Critique of Limits to Growth*, Universe Books, New York, 1973.

31. D. F. Ford and H. W. Dendall, "An Assessment of Emergency Core Cooling Systems—Rule Making Hearing," *Union of Concerned Scientists*, April 1973.

32. T. B. Cochran, *The Liquid Metal Fast Breeder Reactor: An Economic and Environmental Critique*, Johns Hopkins Univ. Press, Baltimore, 1974.

33. A. B. Lovins and J. H. Price, *Non Nuclear Futures*, Friends of the Earth, New York, 1975.

34. A. V. Kneese, "The Faustian Bargain," *Resources*, **44** (1) (September 1973).

35. M. Willrich and T. B. Taylor, *Nuclear Theft: Risks and Safeguards*, Ballinger, Cambridge, MA, 1974.

36. M. Willrich (ed.), *International Safeguards and Nuclear Industry*, Johns Hopkins University Press, Baltimore, 1973.

37. W. B. Bader, *The United States and the Spread of Nuclear Weapons*, Pegasus, New York, 1968.

38. R. U. Ayres, "Technology and Materials Substitution," presented at the 1976 Joint Annual Meeting of the American Institute of Physics and the American Physical Society, New York, February 2–5, 1976.

39. United Nations, "Standard International Trade Classifications," UN Economic and Social Council E/CN 3/456, New York, May 1974.

40. United Nations, "International Standard Classification of All Goods and Services," UN Economic and Social Council, E/CN 3/457, New York, June 1974.

41. R. U. Ayres, "A Materials–Process–Product Model," in A. V. Kneese and B. Bower (eds.), *Environmental Quality and the Social Sciences: Theoretical and Methodological Studies*, Johns Hopkins University Press, Baltimore, 1972.

42. R. U. Ayres and S. Noble, "Materials Scarcity and Substitution," IRT-302-R, prepared for The Conference Board, New York, October 1972.

43. J. J. Siirola and D. F. Rudd, "Computer-aided Synthesis of Chemical Process Networks—From Reaction Path Data to the Process Task Network," *Indus. Eng. Chem. Fund.*, **10** (3) (1971).

44. V. Pareto, *Manuel d'Economie Politique*, Giard et Briere, Paris, 1909.

45. A. C. Pigou, *Economics of Welfare*, Macmillan, New York, 1952.

46. E. K. Brandt and D. O. Wood, "An Economic Interpretation of the Energy/GNP Ratio," in M. Makrakis (ed.), *Energy: Demand, Conservation and Institutional Problems*, MIT Press, Cambridge, MA, 1973.

47. N. Georgescu-Roegen, *South. Econ. J.*, **41**, 347 (1975).

48. M. K. Hubbert, *Sci. Amer.*, **225**, 60 (September 1971); also, W. E. Ricker, in National Academy of Science, National Research Council, *Resources and Man*, Freeman, San Francisco, 1961.

49. R. S. Berry and M. F. Fels, *Bull. Atom. Sci.*, **29**, 11 (1973).

50. C. Bullard and R. Herendeen, "Energy Cost of Goods and Services, 1963 and 1967," Center for Advanced Computation, University of Illinois, Urbana, IL, 1973.

51. B. Hannon, "System Energy and Recycling: A Study of the Beverage Industry," Center for Advanced Computation, University of Illinois, Urbana, IL, 1972.

52. K. Watt, "Land Use, Energy Flow and Decision-Making in Human Society," University of California, Davis, 1973.

53. D. A. Huettner, "Net Energy Analysis: An Economic Assessment," *Science*, **192**, 101–104 (April 1976).

APPENDIX 3-A

PROPERTIES APPLICABLE TO DURABLE MATERIALS

Incompressability. Resistance to compression, that is, low coefficient of dimensional change along the direction of a compressional stress (static).

Inertia. Massiveness; resistance to being moved by application of external force.

Rigidity (inflexibility, stiffness). Resistance to dimensional change or distortion under any *static* stress–strain combination (compression, tension, bending, torsion). Resistance to forces tending to change shape.

Strength. Resistance to *dynamic* stress–strain combinations; shatter-resistance, shock-resistance, endurance.

Tensile strength. Resistance to strain, that is, low coefficient of dimensional change along direction of tensile strain (static and dynamic).

Toughness. Resistance to "fatigue," that is, low rate of decrease of strength as a function of time and repeated stresses and strains.

Flexibility. Lack of rigidity with respect to dynamic stress–strain in one or two internal dimensions but not all three dimensions. Thus a woven fabric is resistant to changes in dimension in two internal dimensions; but free to change in the third dimension by folding. A single fiber tends to be resistant to change (stretching) in only one dimension.

Elasticity. Flexibility with respect to dimensional change or distortion under static stress and/or strain, plus a tendency to return to the original dimensional form (shape) when stresses and strains are relaxed, as long as the stress–strain is below the so-called "elastic limit." Rubber is an exemplary elastic substance and "rubbery" is a synonym for elastic.

Plasticity. Flexibility without elasticity (i.e., the elastic limit is low) where stress–strain forces tend to produce permanent distortions.

Weather resistance. Ability to retain useful mechanical properties despite surface exposure to air (oxygen), water vapor, and various "normal" contaminants of air and water such as NaCl, SO_x, NO_x, CO_2, and their ionic forms.

Temperature resistance. Ability to retain useful mechanical properties despite exposure to elevated temperatures (as well as air and water vapor).

Impenetrability. Resistance to intrusion or penetration by foreign objects (e.g., projectiles), plus impermeability to fluids.

Absorbency. Permeability to fluids, especially liquids.

Adhesion. Ability to form a strong bond with dissimilar materials. Cement is an exemplary adhesive substance.

PROPERTIES APPLICABLE TO CONSUMABLE MATERIALS

Surface activity. The ability to cover a surface and alter its properties significantly. Surfactants achieve their effect through altering the behavior of dirt particles, by "wetting" them, which breaks the bond between the dirt and the surface of the object to be cleaned, then suspends the particle in the surrounding fluid (usually water). Flocculants, in turn, cause the tiny suspended particles to cling to each other and form larger clumps that are relatively easy to remove by filtration. Soap is a kind of flocculant, combining physically with dirt particles and causing them to adhere together rather than to the object being washed.

Solvent power. The ability to dissolve a solid by physically breaking the bonds between individual molecules of solute, or by causing the molecules of solute themselves to ionize.

Viscosity. The internal resistance of a fluid to forced motion. A low viscosity fluid is one that flows easily and vice versa.

APPENDIX 3-B

ELEMENTARY MICROPROCESSES

P_1 Transportation (solids or fluids)
- $P_{1.1}$ Vehicle, self-propelled (truck)
  - (Solid or containerized fluid)
- $P_{1.2}$ Vehicle, passive (pallet; cable car, or pneumatic tube
  - (Solid or containerized fluid)
- $P_{1.3}$ Conveyor belt
  - (Solid or containerized fluid)
- $P_{1.4}$ Gravity flow (channel or pipe)   (Fluid)
- $P_{1.5}$ Pressure flow (pump or pipe)   (Fluid)
- $P_{1.6}$ Convective flow   (Fluid)
- $P_{1.7}$ Capillary flow   (Fluid)

P_2 Change of energy state (solid or fluid)
- $P_{2.1}$ Irradiation
- $P_{2.2}$ Electrification or magnetization
- $P_{2.3}$ Heating
  - $P_{2.3.1}$ No change of phase
  - $P_{2.3.2}$ Change of phase
- $P_{2.4}$ Refrigeration
  - $P_{2.4.1}$ No change of phase
  - $P_{2.4.2}$ Change of phase

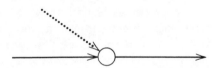

P_3 Change in physical form (mainly solids)
- $P_{3.1}$ Pressure forming (or compression, for fluids)
- $P_{3.2}$ Extension or expansion (or evacuation, for gases)
- $P_{3.3}$ Torsion (twist)
- $P_{3.4}$ Shear (bend)
- $P_{3.5}$ Alignment of fibers or filaments (carding, comb)
- $P_{3.6}$ Winding of fibers or filaments (cone, quill, spool)
- $P_{3.7}$ Randomization of fibers or filaments (felting)

P_4 Physical integration (solids or mesh)
- $P_{4.1}$ Fusion or sintering
- $P_{4.2}$ Adhesion
- $P_{4.3}$ Weaving of filaments (spinning, braiding, roving)
- $P_{4.4}$ Mechanical joining (sew, rivet, screw, bolt, nail)
- $P_{4.5}$ Implantation in a surrounding medium

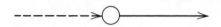

P_5 Physical disintegration (solids or mesh)
- $P_{5.1}$ Shock
  - $P_{5.1.1}$ Mechanical impact (chopping, splitting)
  - $P_{5.1.2}$ Explosive impact (blasting)
  - $P_{5.1.3}$ Acoustic
- $P_{5.2}$ Tooth cutting (sawing, slicing, drilling, milling)
- $P_{5.3}$ Crushing
- $P_{5.4}$ Tearing or picking

P_6 Physical association
- $P_{6.1}$ Mechanical stirring or blending
- $P_{6.2}$ Mixing by acoustic agitation
- $P_{6.3}$ Entrainment and suspension by moving fluid stream
- $P_{6.4}$ Solution
- $P_{6.5}$ Absorption
- $P_{6.6}$ Adsorption
- $P_{6.7}$ Electrostatic deposition
- $P_{6.8}$ Diffusion

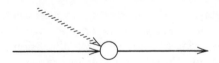

P_7 Physical dissociation or separation
- $P_{7.1}$ Mechanical dismantling
- $P_{7.2}$ Sifting and sorting
- $P_{7.3}$ Filtration
- $P_{7.4}$ Centrifugal separation
- $P_{7.5}$ Flocculation/precipitation
- $P_{7.6}$ Settle/drain
- $P_{7.7}$ Crystallization

$P_{7.8}$ Evaporation, melting, or sublimation (see $P_{2.3.2}$ or $P_{3.2}$)

$P_{7.9}$ Condensation or freezing (see $P_{2.4.3}$ or $P_{3.1}$)

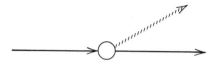

P_8 Surface treatment or finishing (solids)

    $P_{8.1}$ Surface removal

        $P_{8.1.1}$ Abrasion (grinding, polishing, sanding, napping)

        $P_{8.1.2}$ Etching

        $P_{8.1.3}$ Coating removal (pickling, scaling, bleaching, stripping)

    $P_{8.2}$ Surface nonadditive treatment

        $P_{8.2.1}$ Work hardening

        $P_{8.2.2}$ Heat treatment

        $P_{8.2.3}$ Pressing

    $P_{8.3}$ Surface additive treatment

        $P_{8.3.1}$ Lubrication

        $P_{8.3.2}$ Wetting or fulling (detergent, surfactant)

        $P_{8.3.3}$ Coating (paint, dye, wax, shellac, oil polish, etc.)

        $P_{8.3.4}$ Metal plating

        $P_{8.3.5}$ Reactant coating (anodizing, passivating, nitriding, carbonizing)

P_9 Chemical dissociation or decomposition

    $P_{9.1}$ Thermal activation (e.g., thermal cracking, dehydration)

    $P_{9.2}$ Electrolytic (e.g., anode reactions)

    $P_{9.3}$ Catalytic intermediary (e.g., catalytic cracking, depolymerization)

    $P_{9.4}$ Hydrolysis (e.g., ions in solution)

    $P_{9.5}$ Photolysis

    $P_{9.6}$ Biological digestion (e.g., proteins → amino acids)

P_{10} Chemical association or synthesis

    $P_{10.1}$ Thermal activation (e.g., combustion)

    $P_{10.2}$ Electrolytic (e.g., cathode reactions)

    $P_{10.3}$ Catalytic intermediary (e.g., polymerization, condensation, isomerization)

    $P_{10.4}$ Hydration

    $P_{10.5}$ Photochemical reaction (e.g., smog)

    $P_{10.6}$ Biological synthesis (e.g., amino acids → proteins)

Table 3-B.1 Processes Applicable to Simple Material Forms

		P2						P3							P4					P5				P8										P9				P11
		.1	.2	.3.1	.3.2	.4.1	.4.2	.1	.2	.3	.4	.5	.6	.7	.1	.2	.3	.4	.5	.1	.2	.3	.4	.1.1	.1.2	.2.1	.2.2	.2.3	.3.1	.3.2	.3.3	.3.4	.3.5	.1	.3	.5	.6	.1
M1 metal or alloy (conductor)	S	X	X	X	X			X	X	X	X				X	X	X	X	X	X	X	X			X	X	X		X	X	X	X	X					
	G	X	X	X	X			X	X	X	X				X	X	X	X	X	X	X	X			X	X	X		X	X	X	X	X					
	F	X	X	X	X			X	X	X	X	X	X	X	X	X	X	X	X	X	X	X			X	X	X		X	X	X	X	X					
	M	X	X	X	X		X	X	X	X	X	X			X	X	X	X	X	X	X	X	X		X	X	X		X	X	X	X	X					
	L	X	X	X	X				X						X	X	X	X	X	X	X	X	X															
M2 polar electrolytes soluble in H2O	S	X	X	X	X			X	X	X					X	?	X	X	X			X	X											X	X			
	G	X	X	X	X			X	X	X					X	?	X	X	X			X	X											X	X			
	L	X	X	X	X																													X	X		X	X
	V	X	X	X	X	X			X																									X	X	?	X	X
M3 polar non-electrolytes soluble in H2O	S	X	X	X	X			X	X	X					X	?	X	X	X			X	X											X	X			
	G	X	X	X	X			X	X	X					X	?	X	X	X			X	X											X	X	?	X	X
	L	X	X	X	X	X																												X	X	?	X	X
	V	X	X	X	X	X			X																									X	X	?		
M4 nonpolar compounds soluble in nonpolar	S	X	X	X	X			X	X	X	X				X	X	X	X	X	X	X	X		?	X	X	X		X	X	X	X	X					
	G	X	X	X	X			X	X	X	X				X	X	X		X	X	X	X			X	X	X		X	X	X	X	X					X
	F	X	X	X	X			X	X	X	X	X	X	X	X	X	X	X	X	X	X	X			X	X	X		X	X	X	X	X					X
	M	X	X	X	X			X	X	X	X	X			X	X	X	X	X	X	X	X			X	X	X		X	X	X	X	X					
	L	X	X	X	X	X			X						X	X	X	X	X	X	X	X	X											X	X		X	X
	V	X	X	X	X	X			X																									X	X		X	X
M5 insoluble organic	S	X	X	X	X			X	X	X	X				X	X	X	X	X	X	X	X		?	X	X	X		X	X	X	X	X					
	G	X	X	X	X			X	X	X	X				X	X	X		X	X	X	X			X	X	X		X	X	X	X	X					
	F	X	X	X	X			X	X	X	X	X	X	X	X	X	X	X	X	X	X	X			X	X	X		X	X	X	X	X					X
	M	X	X	X	X			X	X	X	X	X			X	X	X	X	X	X	X	X			X	X	X		X	X	X	X	X					
	L	X	X	X	X	X			X						X	X	X	X	X	X	X	X	X															
M6 insoluble inorganic compounds	S	X	X	X	X			X	X	X	X				X	X	X	X	X	X	X	X		?	X	X	X		X	X	X							
	G	X	X	X	X			X	X	X	X				X	X	X		X	X	X	X			X	X	X		X	X	X			X	X	X	X	
	F	X	X	X	X			X	X	X	X	X	X	X	X	X	X	X	X	X	X	X			X	X	X		X	X	X			X	X	X	X	
	M	X	?	X	X			X	X	X	X	X			X	X	X	X	X	X	X	X																
	L	X	X	X	X	X			X																									X	X	X	X	
	V	X	X	X	X	X			X																									X	X	X		

74

Table 3-B.2 Processes Affecting Composite Materials

	P6								P7									P8	P9		P10					
	.1	.2	.3	.4	.5	.6	.7	.8	.1	.2	.3	.4	.5	.6	.7	.8	.9	.1.3	.2	.4	.1	.2	.3	.4	.5	.6
S–S								X								X		X								
G–S						X										X					X					
F–S																X					X					
M–S								X								X					X					
L–S				X	X									X		X			X		X	X	X	X		
V–S				X	X	X											X		X		X	X	X	X		
G–G	X	X				X				X	X					X					X					
F–G = G–F	X	X								X	X					X					X					
L–G				X	X						?		X			X			X		X	X	X	X		
V–G				X	X	X										X			X		X	X	X	X		
F–F	X	X								X	X					X					X					
G–M											X					X					X					
L–M						X					X		X			X			X		X	X	X	X		
V–M						X											X		X		X	X	X	X		
G–L[a]	X	X	X	X			X			X	X	X	X		X[a]	X			X	X	X	X	X	X	X	X
F–L	X	X	X	X						X	X	X	X			X			X		X	X	X	X	X	
L–L	X	X		X			X			X		X		X		X					X			?	X	
V–L		X		X												X		X			X			X	X	
G–V		X	X				X			X	X		X			X			X		X	X	X	X	X	
F–V		X	X							X	X		X			X			X		X	X	X	X	X	
L–V		X									X		X			X					X			X	X	
V–V							X									X					X			X	X	

[a]Includes solutions—by convention, a solution of polar or nonpolar solid material in a liquid is classed as a granule–liquid (G–L) composite (rather than a solid–liquid or filament–liquid).

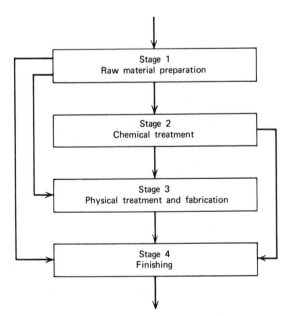

Figure 3-B.1. Generalized process diagram.

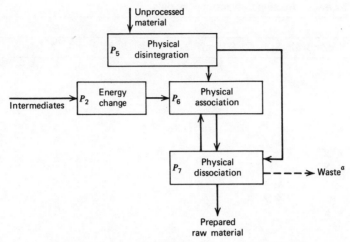

Figure 3-B.2. Stage 1 flow diagram: Raw material preparation. *a*—Waste streams are generally associated with P_7 (physical dissociation) processes simply because the waste must be *separated* from the useable materials by such a process. Losses can, of course, occur at any point in the system.

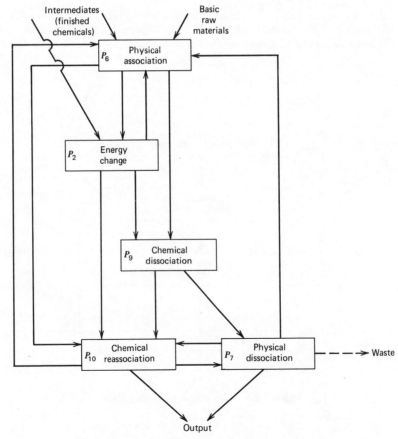

Figure 3-B.3. Stage 2 flow diagram: Chemical treatment.

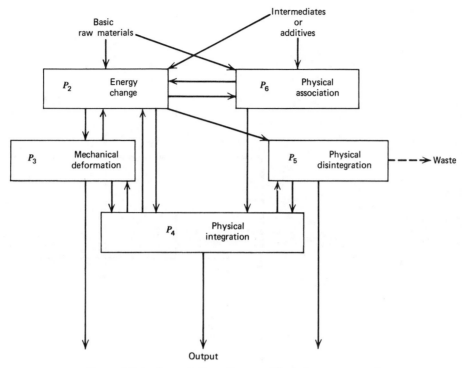

Figure 3-B.4. Stage 3 flow diagram: Physical treatment and fabrication.

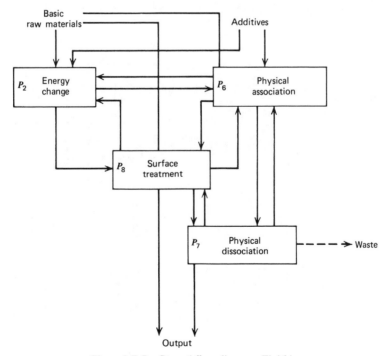

Figure 3-B.5. Stage 4 flow diagram: Finishing.

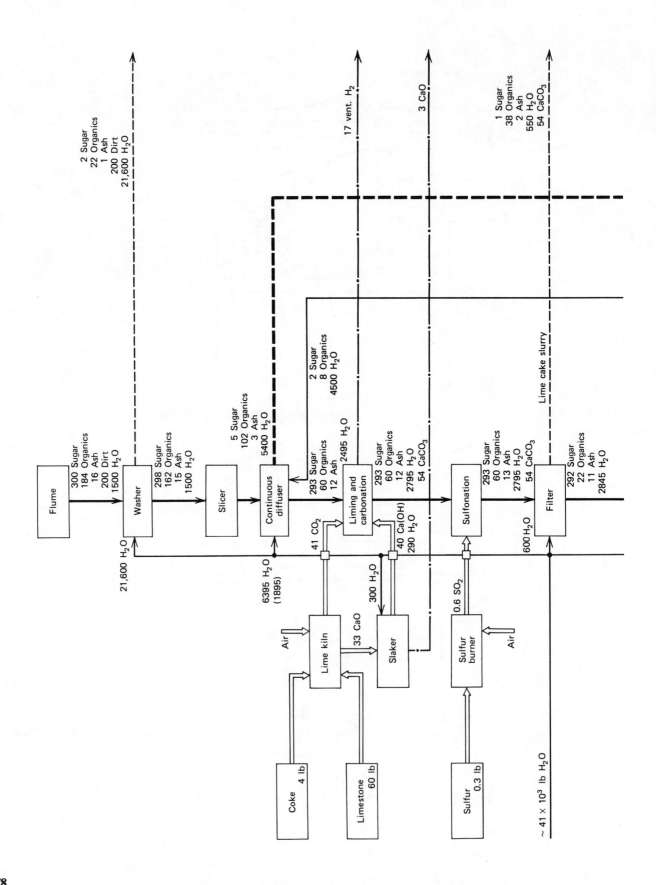

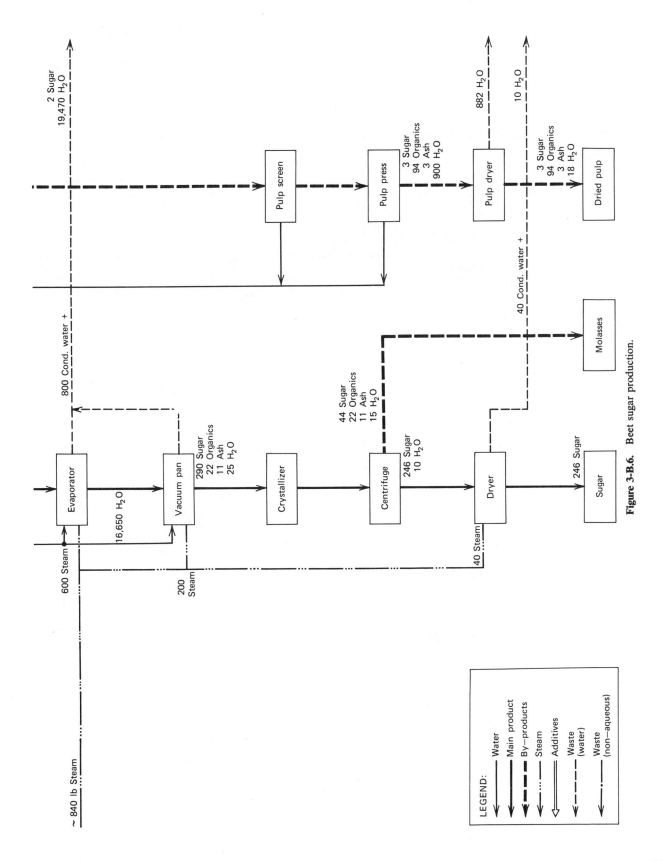

Figure 3-B.6. Beet sugar production.

79

BEET SUGAR	1 Washer	2 Slicer	3 Continuous Diffuser	4 Lime Kiln	5 Slaker	6 Liming & Carbonation	7 Sulfur Burner
BEETS — S=SUGAR	300						
O=ORGANICS	184						
A=ASH & DIRT	216						
COKE = C				4			
CaO				19			
CO₂				41			
Sulfur							0.3
Oxygen = O₂				10.7			0.3
H₂O	1,500 / 21,800		63.95		300		
1 Washer		1 1 1 / 1					
2 Slicer			.99 .95 1 / .25				
3 Continuous Diffuser						1 1 1 / .89	
4 Lime Kiln					1		1
5 Slaker						1 / .11	
6 Liming & Carbonation							
7 Sulfur Burner							
8 Sulfonation							
9 Filter							
10 Evaporator							
11 Vacuum Pan							
12 Crystallizer							
13 Centrifuge							
14 Dryer							
15 Pulp Screen							
16 Pulp Press			.01 .05 / .75				
17 Pulp Dryer							

KEY:

SUGAR	ORGANIC	ASH ETC.
COKE	CaO	CO₂
SULFUR	O₂	H₂O

Figure 3-B.7.

8 Sulfonation	9 Filter	10 Evaporator	11 Vacuum Pan	12 Crystallizer	13 Centrifuge	14 Dryer	15 Pulp Screen	16 Pulp Press	17 Pulp Dryer	PRODUCT	WASTE
	600	16.650 / 600 STEAM	2.00 STEAM			40 STEAM					
											2·S / 22·0 / 201·dirt / 21,600 H₂O
										3 CaO	
1 1 1 / 1 1 / 1											17 H₂
1 1											
	1 1 1 / 1 1 / 1										
		1 1 1 / 1								I S 38 0 / 2·A,54·CaCO₃ / 550 H₂O
			1 1 1 / 1							2 S / 800·H₂O (L) / 19,270·H₂O (V)
				1 1 1 / 1						200 H₂O (V)
					1 1 1 / 1					
						1 / 1				Molasses / 44·S, 22·0	
										Sugar / 246	40 H₂O (L) / 10 H₂O (V)
							1 1 1 / 1				
								1 1 1 / 1			
										Dried Pulp / 3S, 94·0 / 3A, 18·H₂O	

Figure 3-B.7. (Continued)

BEET SUGAR

	1 Washer	2 Slicer	3 Continuous Diffuser	4 Lime Kiln	5 Slaker	6 Liming & Carbonation	7 Sulfur Burner	8 Sulfonation
BEETS S=SUGAR	300							
O=ORGANICS	184							
A=ASH & DIRT	216							
COKE = C				4				
CaO				19				
CO_2				41				
Sulfur							0.3	
Oxygen = O_2				10.7			0.3	
H_2O	1500 / 21,600		63.95		300			
1 Washer		298·S 162·O 15·A 1500 H_2O						
2 Slicer			298 S 162 O 15 A 1500 H_2O					
3 Continuous Diffuser						292 S 60 O 12 A 2495 H_2O		
4 Lime Kiln					33 CaO	41 CO_2		
5 Slaker						40 Ca(OH)₂ (·21 CaO·19 H_2O) 290 H_2O		
6 Liming & Carbonation								293 S 60·0.12A 54 C_aCO_3 2795 H_2O
7 Sulfur Burner								0.6 SO_2 (0.3 O, 0.3 S)
8 Sulfonation								
9 Filter								
10 Evaporator								
11 Vacuum Pan								
12 Crystallizer								
13 Centrifuge								
14 Dryer								
15 Pulp Screen								
16 Pulp Press			2·S, 8·O 4500 H_2O					
17 Pulp Dryer								

Figure 3-B.8.

9 Filter	10 Evaporator	11 Vacuum Pan	12 Crystallizer	13 Centrifuge	14 Dryer	15 Pulp Screen	16 Pulp Press	17 Pulp Dryer	PRODUCT	WASTE
←600	16,650→ 600 STEAM	200 STEAM			40 STEAM					
										2·S 22·0 201·dirt 21,600 H$_2$O
					5S , 102·0 5·A 5400 H$_2$O					
										3 CaO
										17 H$_2$
293·S,60·0 13·A,54·CaCO$_3$ (17 CaO·37 CO$_2$) 2795·H$_2$O										
	292·S, 22·0 11·A 2845 H$_2$O									1 S 380 2·A,54·CaCO$_3$ 550 H$_2$O
		292·S, 22·0 11·A 225 H$_2$O								2 S 800·H$_2$O (L) 19270·H$_2$O (V)
			290·S,22·0 11 A 25 H$_2$O							200 H$_2$O (V)
				290·S , 22·0 11·A 25 H$_2$O						
					246 S 10 H$_2$O				Molasses 44·S , 22·0	
									Sugar 246	40 H$_2$O (L) 10 H$_2$O (V)
							5·S , 102·0 3A 5400 H$_2$O			
								3·S , 94·0 3A 900 H$_2$O		
									Dried Pulp 3·S , 94·0 3A , 18·H$_2$O	

Figure 3-B.8. (Continued)

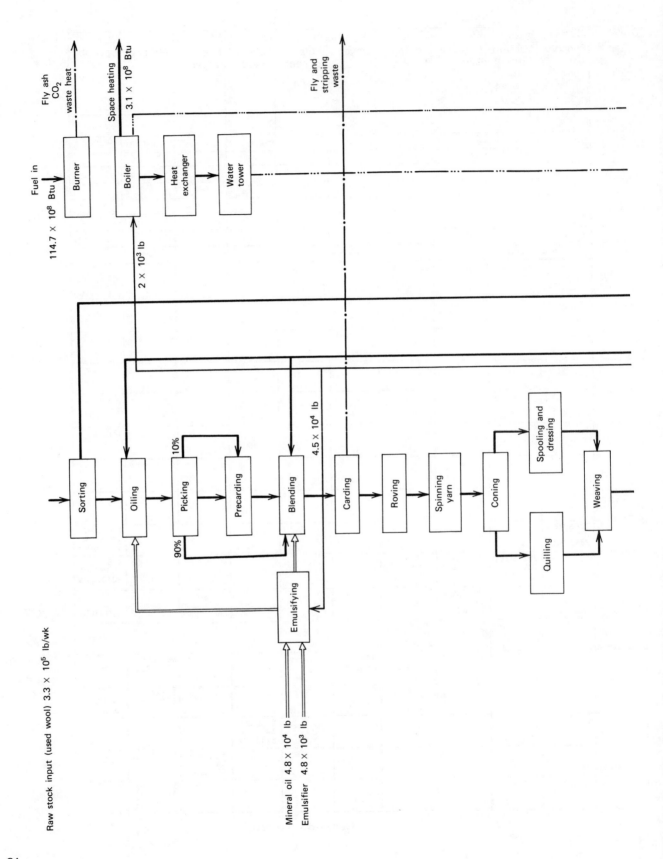

Raw stock input (used wool) 3.3×10^5 lb/wk

Mineral oil 4.8×10^4 lb
Emulsifier 4.8×10^3 lb

Fuel in
114.7×10^8 Btu

Fly ash
CO_2
waste heat

Space heating
3.1×10^8 Btu

Fly and
stripping
waste

2×10^3 lb

4.5×10^4 lb

Burner

Boiler

Heat
exchanger

Water
tower

Sorting

Oiling

Picking

10%

90%

Precarding

Blending

Carding

Roving

Spinning
yarn

Coning

Spooling and
dressing

Quilling

Weaving

Emulsifying

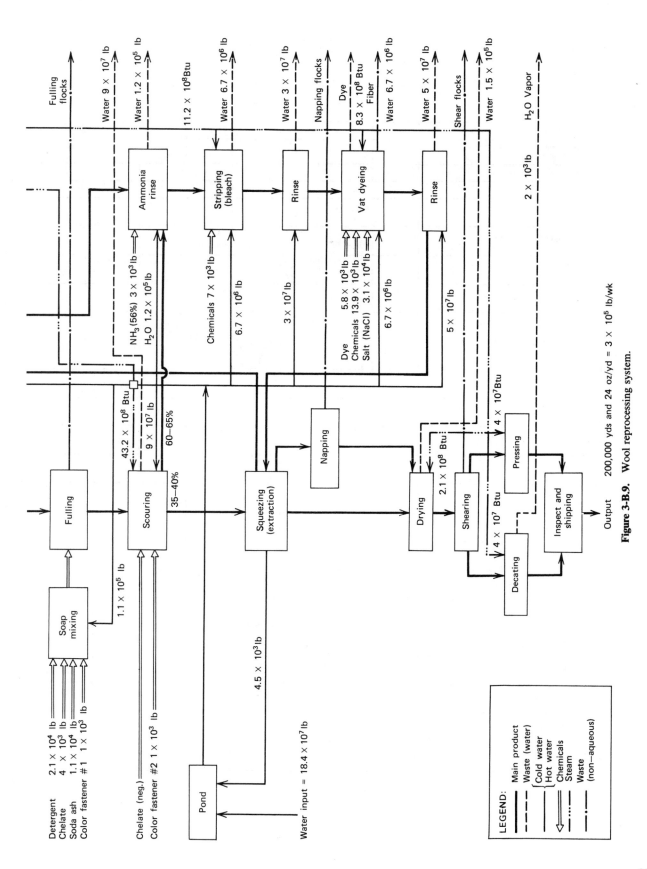

Figure 3-B.9. Wool reprocessing system.

LEGEND:
— Main product
– – – Waste (water)
{ ⇒ Cold water / Hot water
⇒ Chemicals
⇒ Steam
⋯⋯ Waste
–·– (non—aqueous)

Detergent 2.1 × 10⁴ lb
Chelate 4 × 10³ lb
Soda ash 1.1 × 10⁴ lb
Color fastener #1 1 × 10³ lb

Chelate (neg.)
Color fastener #2 1 × 10³ lb

Water input = 18.4 × 10⁷ lb

1.1 × 10⁵ lb

4.5 × 10³ lb

Soap mixing

Fulling

Scouring

43.2 × 10⁸ Btu
9 × 10⁷ lb
60—65%
35—40%

Pond

Squeezing (extraction)

Napping

Drying

Shearing

Decating

Pressing

2.1 × 10⁸ Btu
4 × 10⁷ Btu
4 × 10⁷ Btu

Inspect and shipping

4 × 10⁷ Btu

Output

200,000 yds and 24 oz/yd = 3 × 10⁵ lb/wk

Fulling flocks
Water 9 × 10⁷ lb
Water 1.2 × 10⁵ lb
11.2 × 10⁸ Btu
Water 6.7 × 10⁶ lb
Water 3 × 10⁷ lb
Napping flocks
Dye
8.3 × 10⁸ Btu
Fiber
Water 6.7 × 10⁶ lb
Water 5 × 10⁷ lb
Shear flocks
Water 1.5 × 10⁵ lb
H₂O Vapor

Ammonia rinse

Stripping (bleach)

Rinse

Vat dyeing

Rinse

NH₃ (56%) 3 × 10³ lb
H₂O 1.2 × 10⁵ lb
Chemicals 7 × 10³ lb
6.7 × 10⁶ lb
3 × 10⁷ lb

Dye 5.8 × 10³ lb
Chemicals 13.9 × 10³ lb
Salt (NaCl) 3.1 × 10⁴ lb
6.7 × 10⁶ lb

5 × 10⁷ lb

2 × 10³ lb

WOOL REPROCESS	1 Sort	2 Emulsify	3 Oil	4 Pick	5 Precard Card	6 Blend	7 Rove Spin Weave	8 Cone Quill Spool	9 Mix Soap	10 Full
WOOL (F)	330,000									
OIL & EMULSIFIERS		52,800								
DETERGENT & CHELATE									21,000 det. 5,000 chel.& CF 11,000 soda ash	
AMMONIA & STRIPPING CHEMICALS										
DYE & CHEMICALS										
WATER (H_2O)		45,000							110,000	
1 Sort (7.1)			300,000 F							
2 Emulsify (6.3.1)			50,400 Oil 43,000 H_2O			2,400 Oil 2,000 H_2O				
3 Oil (9.3.1)				315,000 F 50,400 Oil 58,000 H_2O						
4 Pick (5.4)					31,500 F 5,040 Oil 5,800 H_2O	283,500 F 45,360 Oil 52,200 H_2O				
5 Precard Card (3.5)						31,500 F 5,040 Oil 5,800 H_2O	315,000 F 50,400 Oil 71,500 H_2O			
6 Blend (6.1)					330,000 F 52,800 Oil 75,000 H_2O					
7 Rove, Spin Weave (4.3)								315,000 F 50,400 Oil 71,500 H_2O		315,000 F 50,400 Oil 71,500 H_2O
8 Cone, Quill Spool (3.6)							315,000 F 50,400 Oil 71,500 H_2O			
9 Mix, Soap (6.2.1)										37,000 Soap 110,000 H_2O
10 Full (9.3.2)										
11 Scour (8.1)										
12 Ammonia Rinse (8.3)										
13 Strip (9.1.3)										
14 Rinse (8.3)										
15 Vat Dye (9.3.3)										
16 Extract (7.2.3)			15,000 F 15,000 H_2O			15,000 F 15,000 H_2O				
17 Nap Shear (9.1.1)										
18 Dry (7.3.1)										
19 Decate (9.2.2)										
20 Press (9.2.3)										

Figure 3-B.10.

11 Scour	12 Ammonia Rinse	13 Strip	14 Rinse	15 Vat Dye	16 Extract	17 Nap. Shear	18 Dry	19 Decate	20 Press	PRODUCT	WASTE
	3,000 NH₃	7,000 Chem									
				5800 dye 44,900 chem							
90,000,000	120,000	6,700,000	80,000,000	6,700,000	-150,000			2,000 steam			
	30,000 W										
											1,500 F 2400 Oil 3,500 H₂O
											3000 F 400 Oil 1500 H₂O
312,000 F 50,000 Oil 37,000 Soap 180,000 H₂O											
	208,000 F 200,000 H₂O				104,000 F 100,000 H₂O						50,000 Oil 37,000 Soap 89880,000 H₂O
		208,000 F 200,000 H₂O									3,000 NH₃ 120,000 H₂O
			208,000 F 200,000 H₂O								7,000 Chem 6,700,000 H₂O
				208,000 F 200,000 H₂O							30,000,000 H₂O 50,000,000 H₂O
			213,000 F 200,000 H₂O		213,000 F 200,000 H₂O						3,000 F 800 Dye 44,900 Chem 6700,000 H₂O
						314,000 F 150,000 H₂O					
							309,000 F 150,000 H₂O				5,000 F Nap 9,000 F Shear
						309,000 F		30,000 F	270,000 F		150,000 H₂O (V)
										30,000	2,000 H₂O (V)
										270,000	

Figure 3-B.10. (Continued)

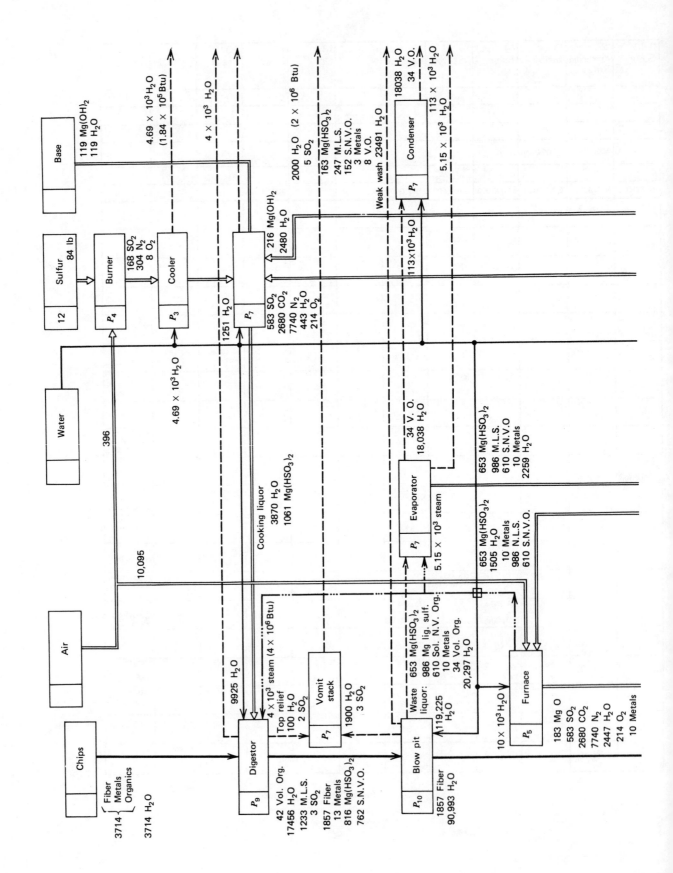

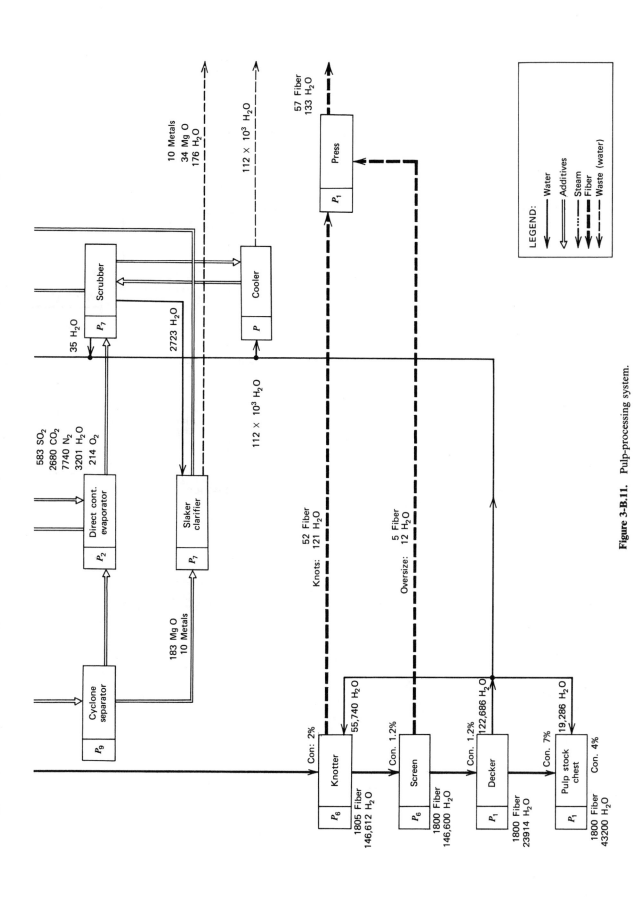

Figure 3-B.11. Pulp-processing system.

89

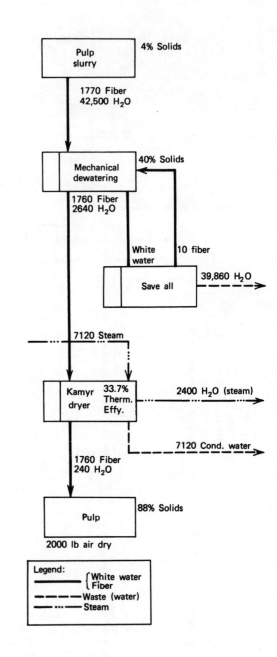

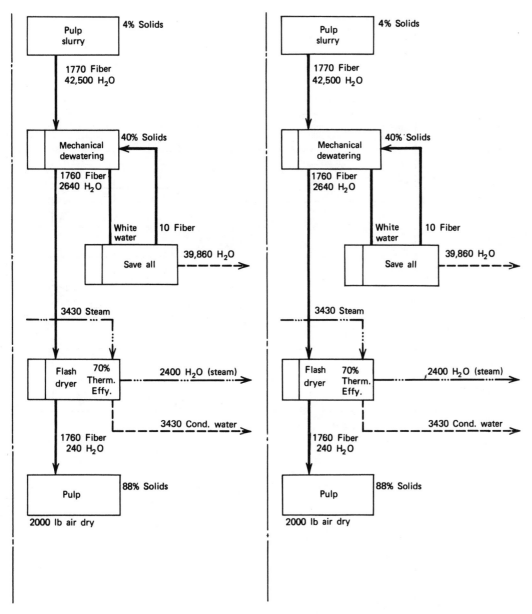

Figure 3-B.12. Pulp drying.

	1 Sulfur Burner	2 Absorbtion Tower	3 Digestor	4 Blow Pit	5 Evaporator 1	6 Evaporator 2	7 Furnace
CHIPS Fiber			1857				
Lignin			1040				
NVSO			762				
Metals& VO			13M + 42 VO				
Sulfur S $_{32}$	84						
MgO $_{59}$		91					
H$_2$O $_{18}$		1398	4000 (S)	119,225	5150 (S)		10,000
O$_2$ $_{32}$	84						2,144
1 Sulfur Burner		168 SO$_2$					
2 Absorbtion Tower			1061 Mg(HSO$_3$)$_2$ 3870 H$_2$O				
3 Digestor				1857 F 1233 MLS 762 NVSO 13M 42 VO 816 Mg(HSO$_3$)$_2$ 17456 H$_2$O			
4 Blow Pit					986 MLS, 610 NVSO, 10M +34VO, 653 Mg(HSO$_3$)$_2$ 20,297 H$_2$O		
5 Evaporator 1						986 MLS 610 NVSO 10M 653 Mg HSO$_3$ 2259 H$_2$O	
6 Evaporator 2							986 MLS 610 NVSO 10M 653 Mg(HSO$_3$) 1505 H$_2$O
7 Furnace							
8 Cyclone Separator						2680 CO 583 SO 2447 H$_2$O	
9 Slaker·Clarifier		216 M OH 2480 H$_2$O					
10 Scrubber		2680 CO$_2$ 583 SO$_2$ 443 H$_2$O					
11 Condenser							
12 Knotter							
13 Screen							
14 Decker							

INTERMEDIATE PROCESS MATERIALS:
S + O$_2$ = SO$_2$
MgO + H$_2$O = Mg(OH)$_2$
Mg(OH)$_2$ + 2 SO$_2$ = Mg(HSO$_3$)$_2$
1040 Lignin + 245 Mg(HSO$_3$)$_2$ = 1233 Mg Lig Sulf + 5 SO$_2$ + 47 H$_2$O

Figure 3-B.13.

8 Cyclone Separator	9 Slaker-Clarifier	10 Scrubber	11 Condenser	12 Knotter	13 Screen	14 Decker	15 Press	FIBER PULP	WASTE
		-35	113,000	55,740		-122,686		19,286	
									Vomit Stack
									5 SO_4 2,000 H_2O (V) 4,000 H_2O (L)
				1857 F 90,993 H_2O					3M·8VO,247MLS 152 NVSO 163 Mg(HSO$_4$) 23,491 H_2O
			34 VO 18 038 H_2O						Weak Wash
		2680 CO_4 583 SO_4 3201 H_2O							
2680 CO, 1OM 583 SO , 183MgO 2447 H_2O									
	1OM 18 MgO								
	2723 H_2O								
									34 VO 131,000 H_2O
					1805F 146 612 H_2O		52 F 121 H_2O		
						1800 F 146,600 H_2O	5 F 12 H_2O		
								1800 F 43,000 H_2O	

Figure 3-B.13. (Continued)

CHAPTER FOUR *Materials/Energy Accounting and Forecasting Models*

with S. B. Noble

The accounting models discussed in this chapter are all essentially designed to help find answers to one class of question, which is always of the form How much material/energy/pollutant "X" is used or generated directly/indirectly in or by process/commodity/product/sector or final demand category "Y"? Such questions are common. Often X refers to total energy in Btu or a particularly scarce fuel such as natural gas or imported petroleum. While Y could be a narrow class of intermediate products—such as chlorine, polyethylene, or aluminum—or a broad sector such as transportation or agriculture. Or Y might refer to a unit of GNP. Obviously X may also stand for a toxic heavy metal such as mercury, when a sudden concern arises as to the true extent of its uses for a variety of purposes. Or X may refer to a specific chemical or family of chemicals such as fluorocarbons or polychlorinated biphenyls (PCBs).

Although the chapter heading does not explicitly so state, the present chapter is essentially concerned with elaborations and applications of the input–output model, whose development has been mainly associated with the name of Wassily Leontief.[†] The usual formulations of the I–O model do not automatically (and usually not in practice) assure that physical requirements, such as the first or

second laws of thermodynamics will be satisfied. But, fortunately, the I–O model is entirely compati-

[†]For those readers interested in historical antecedents, precursors of the input–output model occurred in Russia in the 1920s[1] and in Germany in the early 1930s. Important German writings are by Kaehler[2] and Peter.[3] An interesting American study, similar in significant ways to input–output, is by Loeb.[4] Leontief commented on the Russian developments,[1] participated in the German literature[5] and moved to the United States where he first published on input–output in 1936[6] and 1937.[7] The first edition of his book was published in 1941,[8] the second and enlarged edition, in 1951.[9] He has written many introductory articles on input–output,[10] has edited (and contributed to) a book on research in input–output,[11] and continues to contribute.[12]

The extent of research on input–output is represented by the bibliographies[13, 14] on it and by the many conferences in the field.[15–20] The developments by the U.S. government include the early work in the Bureau of Labor Statistics[21] and by the Air Force.[22, 23] Important examples of large systems of input–output models connected to macro-drivers (these systems are called "closed" input–output models) are by Stone[24] for England, by Wood,[25] Almon et al.,[26] and the U.S. Bureau of Labor Statistics[27] for the United States. The best textbooks on input–output are by Stone and Stone,[28] Miernyk,[29] Leontief,[10] and Chenery and Clark.[30] Most readers find them most accessible in the order given here.

Input–output tables for the United States are provided by the Commerce Department.[31, 32] Projections of input–output tables and of macro-variables in association with them are made by the Bureau of Labor Statistics[33] as the official GNP projections of the U.S. government.

Table 4.1 A Symbolic Representation of the System

				1	2	3	4	5	6	7	8
Opening assets		financial assets	1								
		net tangible assets	2								
Production	Commodities	commodities, basic value	3					$T_{3.5}$	$T_{3.6}$	$T_{3.7}$	$T_{3.8}$
		commodity taxes, net	4					$T_{4.5}$	$T_{4.6}$	$T_{4.7}$	$T_{4.8}$
	Activities	industries	5			$T_{5.3}$	$T_{5.4}$				
		producers of government services	6			$T_{6.3}$					$T_{6.8}$
		private services: domestic service and producers of private n-p services	7			$T_{7.3}$					$T_{7.8}$
Consumption	Expenditure	household goods and services	8								
		government purposes	9								
		purposes of private n-p bodies	10								
	Income and outlay	value added	11			$T_{11.3}$	$T_{11.4}$	$T_{11.5}$	$T_{11.6}$	$T_{11.7}$	
		institutional sector of origin	12								
		form of income	13								
		institutional sector of receipt	14								
Accumulation	Increase in stocks	industries	15								
		producers of government services	16								
	Fixed capital formation	industries	17								
		producers of government services	18								
		producers of private non-profit services to households	19								
	Capital finance	industrial capital formation, land, etc.	20								
		capital transfers	21								
		financial assets	22								
		institutional sectors	23	$T_{23.1}$	$T_{23.2}$						
Rest of the world: current and capital transactions			24	$T_{24.1}$	$T_{24.2}$	$T_{24.3}$			$T_{24.6}$		$T_{24.8}$
Revaluation		financial assets	25								
		net tangible assets	26								
Closing assets		financial assets	27								
		net tangible assets	28								
				1	2	3	4	5	6	7	8

ble with these laws and can be formulated in such a way as to satisfy them rigorously.

What is the relationship between static accounting models and dynamic input–output forecasting models? In the simplest case, the latter are a special application of the former. To convert an input–output type of accounting model into a forecasting model, it is minimally necessary to specify exogenously a future trajectory for final demand and labor force, at a suitable level of disaggregation, keeping everything else constant. The accounting equations are simply exercised again and again for successive years to provide a complete set of time-dependent implications, namely, total production by sector, employment, and so on. Of course, there are a number of ways in which the forecast for later years can be made to reflect what happens in earlier years. The simplest is to introduce expected changes in the input–output coefficients, to reflect changing technologies. A more complex feedback involves adjusting capital spending to reflect demand in earlier years by sector, or to introduce capacity constraints linked to invested capital.

In effect, the dynamic elements are typically introduced by means of a separate econometric model called a "macro-driver," which is mainly an application of statistical analysis of time-series data. Such models are not a primary topic of this book, and I do not intend to comment on their properties or methods of construction.

Section 4.1 briefly describes the national industry–commodity accounts from which I–O tables are developed. This is done to facilitate a later explanation of the differences between industry–industry tables (conventional I–O), industry–commodity tables (as exemplified by the National I–O model of Canada), and commodity–commodity tables, which will be introduced later.

Section 4.2 provides a uniform nomenclature and a concise derivation of the basic input–output equations, including value-added and price relationships. No applications are discussed in this section.

Section 4.3 deals with the first application, to "net energy" analysis. It indicates how I–O tables have been extended through the use of "energy coefficients" to facilitate the approximate determination of indirect (as well as direct) energy consumption associated with a given change in gross production or final consumption levels.

Section 4.4 extends the "dollar" I–O models of

	9	10	11	12	13	14	15	16	17	18	19	20	21	22	23	24	25	26	27	28	
															$T_{1.23}$	$T_{1.24}$					1
															$T_{2.23}$						2
							$T_{3.15}$	$T_{3.16}$	$T_{3.17}$	$T_{3.18}$	$T_{3.19}$					$T_{3.24}$					3
							$T_{4.15}$		$T_{4.17}$	$T_{4.18}$	$T_{4.19}$					$T_{4.24}$					4
																					5
	$T_{6.9}$																				6
		$T_{7.10}$																			7
						$T_{8.14}$										$T_{8.24}$					8
						$T_{9.14}$															9
						$T_{10.14}$															10
															$T_{11.23}$						11
			$T_{12.11}$																		12
				$T_{13.12}$		$T_{13.14}$										$T_{13.24}$					13
			$T_{14.11}$		$T_{14.13}$																14
												$T_{15.20}$									15
															$T_{16.23}$						16
												$T_{17.20}$									17
															$T_{18.23}$						18
															$T_{19.23}$						19
															$T_{20.23}$						20
																					21
															$T_{22.23}$	$T_{22.24}$					22
						$T_{23.14}$							$T_{23.21}$	$T_{23.22}$			$T_{23.25}$	$T_{23.26}$	$T_{23.27}$	$T_{23.28}$	23
				$T_{24.13}$									$T_{24.21}$	$T_{24.22}$		$T_{24.24}$	$T_{24.25}$	$T_{24.26}$	$T_{24.27}$	$T_{24.28}$	24
															$T_{25.23}$	$T_{25.24}$					25
															$T_{26.23}$						26
															$T_{27.23}$	$T_{27.24}$					27
															$T_{28.23}$						28
	9	10	11	12	13	14	15	16	17	18	19	20	21	22	23	24	25	26	27	28	

Section 4.2 to strictly physical units. In particular, an input–output table for the chemical industry is developed. This model can be used to determine changes in "upstream" material/energy requirements or "downstream" price impacts resulting from pollution abatement (or other) changes in the supply or the cost of a given chemical product.

Section 4.5 provides the theoretical framework for extending the I–O formalism to include environmental interactions. The major applications of this elaboration have been to compute industrial residuals generation associated with a given change in production or consumption levels. Methods of assessing abatement costs and optimum abatement levels are also considered.

4.1 INDUSTRY/COMMODITY ACCOUNTS

An accounting model must start from a set of statistical accounts, such as national income and expenditures, and commodity shipments. The current version of the International System of National Accounts is summarized in matrix form in Table 4.1.[34] There are seven major categories and 28 sub-categories in the system. Each row–column intersection $T_{i,j}$ corresponds to a submatrix. Thus submatrix $T_{3.5}$ displays the consumption of commodities, at producers values, by industry sectors. The matrix $T_{5.3}$, on the other hand, shows the output of commodities by industries.

The input–output model in most countries[†] is constructed from two sets of tables derived from the SNA. The first is the so-called *Commodity Table* (Table 4.2),[‡] which is obtained by rearranging the information in rows 5–8 and columns 5–8 of Table 4.1, in fully disaggregated form. The second is a corresponding *Industry Table* (Table 4.3), which is similarly obtained from rows 13–16 and columns 13–16 of Table 4.1.

[†]The United States is an exception to international practice. Input–output tables in the United States are constructed—by a somewhat different procedure—directly from the Census of Manufactures. Tables for the United States economy have been prepared for each year since World War II that there has been a Census of Manufactures: 1947, 1954, 1958, 1963 and 1967. A table for 1972 is now being prepared.

[‡]The term "commodity," here, includes both products and services. The basic UN classification system is the International Classification of Goods and Services (ICGS).[35]

Table 4.2   Sources and Destinations of Commodity Supplies

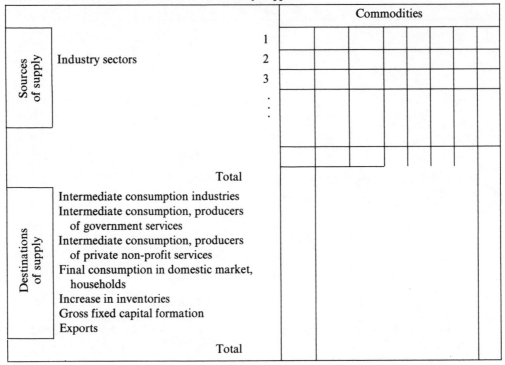

The commodity table displays sources and destinations of all commodities for which accounts are kept. Sources are identified by industry sector, and destinations are recorded by expenditure category, namely, intermediate consumption by industries, government, and households and increase in inventories and capital formation. Each designated commodity is regarded as flowing into or out of an undifferentiated and homogeneous pool, since many sellers do not know who their customers are although they know precisely what commodities they produce and consume. The industry table, in turn, shows the sources (i.e., commodity, labor, and capital inputs) and commodity outputs of each industry. Again, each industry sector is regarded as homogeneous.

The commodity table (Table 4.2) and industry table (Table 4.3) are the raw material for input–output tables—either industry–industry or industry–commodity—although some manipulation is required. It is important to bear in mind that while industries are classified in accordance with their principal (commodity) products, each industrial sector also generates by-products. Correspondingly, commodities are almost invariably produced in

significant quantities by more than a single industry.[†]

A second practical difficulty is that there are normally more commodities than industries, so the industry–commodity input and output matrices are not square, whereas an industry–industry input–output matrix (by definition) must be square; that is, it must have the same number of rows and columns. This involves disaggregation of industries to the commodity level and reaggregation of commodities to the industry level. The details of the procedure need not be described here; suffice it to say that the process involves constructing successive "trial balances" of inputs and outputs until the entire table is balanced. Over the years, the balancing

[†]This difficulty is dealt with in several different ways, depending on the purposes for which the data will be used. The "official" (benchmark) U.S. Input–Output Table, compiled by the Department of Commerce and incorporated in the forecasting model of the Bureau of Labor Statistics (BLS), uses "impure" sectors—in which by-products belonging to different sectors are not separated, but are retained in the same sectors as the primary products. However, most European input–output tables are "purified" in the sense that each sector produces *all* of a given commodity and *only* that commodity. There are technical advantages and disadvantages in each method.

Table 4.3 Industrial Outputs and Costs

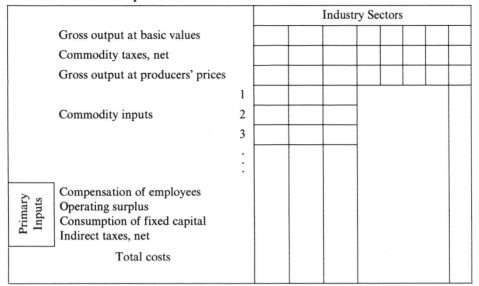

process has given the input–output tables a strong empirical foundation compared to estimates obtained by viewing transactions from only one side (i.e., purchasing or selling).

Another way to avoid the difficulty is to design a model to utilize the industry–commodity account statistics more or less directly. Apart from added mathematical complexity, the model requires an extra market share postulate beyond the standard "industry–technology" assumption of the Leontief model. This somewhat reduces its value for forecasting purposes. Nevertheless, the industry–commodity approach has been followed by one country (Canada). A brief description of the Canadian Industry and Commodity Input–Output Model is attached as an addendum to the next section, although none of the subsequent applications discussed in this chapter utilize it. The Canadian model is a very interesting and important one; it has the advantage of greater detail to compensate for its theoretical drawback.

If the input–output accounts are to be balanced, it is necessary to take into account *all* categories of sales to (or purchases from) the included sectors. For revenue, accounts are introduced that show sales to foreign countries, households, capital purchases, and changes in inventories. For expenditure, accounts are added to record depreciation, taxes and profits. In addition, a distinction is normally introduced between expenditure for purposes

of consumption, by individuals or governments, and expenditure for investment (i.e., for creation of new productive capacity).

4.2 INPUT–OUTPUT MATRICES AND MODELS

When the tedious process of balancing and adjustment of the accounting data is complete, what emerges is a set of transactions, that can be expressed as equations

$$\sum_j T_{ij} + Y_i = X_i \qquad (4.1)$$

where T_{ij} are sales from sector i to sector j, (summed over all purchasing sectors j), Y_i is sales from i to final demand and X_i is total output of sector i. This equation holds whether the indices i, j refer to industry sectors or to commodities (categories of products and services).

The input–output model essentially assumes that the ratio of inputs to total output for any sector is constant over time. In other words, the *input requirements for each sector are assumed to be an unchanging characteristic of the technology of production*. This is known as the industry and technology assumption:

$$\frac{T_{ij}}{X_j} = A_{ij} \qquad (4.2)$$

Thus A_{ij} is the fraction of all inputs to sector j that comes from sector i. The A_{ij} are known as *input coefficients*. Equation (4.1) can then be expressed

$$\sum_j A_{ij}X_j + Y_i = X_i$$

or

$$Y_i = X_i - \sum_j A_{ij}X_j \qquad (4.3)$$

Equation (4.3) can also be expressed more compactly in matrix form

$$Y = (I - A)X \qquad (4.3a)$$

where Y and X are vectors with n elements and A is an $n \times n$ matrix known as the *input* matrix, since its columns represent the fractional inputs to each sector from every other sector.

A comparable but different set of structural relationships might conceivably be postulated by assuming that the allocation of outputs from any sector to all other sectors is constant. This would be tantamount to assuming that each industry always serves exactly the same set of customers—the market share assumption. This is a somewhat less plausible hypothesis than (4.2).[†] Define

$$\frac{T_{ij}}{X_i} = C_{ij} \qquad (4.4)$$

Thus C_{ij} is the fraction of outputs of sector i that goes to sector j, or the *output coefficient*. In this case, Equation (4.1) becomes

$$\sum_j C_{ij}X_i + Y_i = X_i \qquad (4.5)$$

In matrix form,

$$Y = X(I - C) \qquad (4.5a)$$

Note that by equating (4.3a) and (4.5a) we have

$$X(I - C) = (I - A)X$$

Whence, premultiplying both sides by X^{-1} one obtains

$$C = I - X^{-1}(I - A)X \qquad (4.6)$$

It is clear from (4.6) that if C_{ij} were constant, then A_{ij} could not be constant, and vice versa. Thus the two possible input–output hypotheses are mutually exclusive.[†] However, the output coefficients, C_{ij}—while *not* assumed to be constant over time —are quite helpful in certain types of analysis, as will be seen later.

Equation (4.3a) can be formally solved by premultiplying both sides by $(I - A)^{-1}$, yielding the result:

$$X = (I - A)^{-1}Y = BY \qquad (4.7)$$

Similarly, (4.5a) can be formally solved by postmultiplication of both sides by the matrix $(I - C)$, giving

$$X = Y(I - C)^{-1} = YD \qquad (4.8)$$

It is convenient to define an inverse *input* matrix B, satisfying the identities:

$$B = (I - A)^{-1}$$
$$= I + AB$$
$$= I + A + A^2B$$
$$= I + A + A^2 + A^3 + \cdots \qquad (4.9)$$

Similarly, it is convenient to define an inverse *output* matrix D as follows:

$$D = (I - C)^{-1}$$
$$= I + CD$$
$$= I + C + C^2D$$
$$= I + C + C^2 + C^3 + \cdots \qquad (4.10)$$

These algebraic relationships are convenient for certain kinds of practical analysis, as will now be shown. Notice that

$$[A^2]_{ij} = \sum_k A_{ik}A_{kj}$$

$$[A^3]_{ij} = \sum_{k,l} A_{ik}A_{kl}A_{lj}$$

and similar expressions can be written for $[C^2]_{ij}$, $[C^3]_{ij}$, and the like. Substituting the sums into (4.9)

[†]This assumption is made, however, in the Canadian Industry–Commodity Model summarized briefly at the end of this section.

[†]This is true in the case of a square input–output matrix.

or (4.10) gives the following

$$B_{ij} = \delta_{ij} + A_{ij} + \sum_k A_{ik}A_{kj} + \sum_{k,l} A_{ik}A_{kl}A_{lj} + \cdots$$

(4.11)

$$D_{ij} = \delta_{ij} + C_{ij} + \sum_k C_{ik}C_{kj} + \sum_{k,l} C_{ik}C_{kl}C_{lj} + \cdots$$

(4.12)

where δ_{ij} is the so-called Kronecker δ-function,

$$\delta_{ij} = \begin{cases} 1 & i = j \\ 0 & i \neq j \end{cases}$$

In words, the inverse input matrix element B_{ij} consists of sums of the "direct" terms plus all nonzero pair-wise path-products, plus the sum of all nonzero three-link path-products, and so on. Each path-product represents a chain of real transactions—that is, flows of goods and services—in the economy. The inverse output matrix element can also be expressed as a sum of two-step, three-step, and multiple-step path-product of output coefficients.

To understand these chains of transactions more clearly and interpret their significance, let us consider a small set of intersectoral flows as illustrated in Figure 4.1. Each motor vehicle utilizes a certain amount of steel. Some is purchased directly (path $5 \rightarrow 2$), while some is incorporated in stampings (path $5 \rightarrow 4 \rightarrow 2$), and some in motor vehicle parts (path $5 \rightarrow 3 \rightarrow 2$). There are also many other pathways, not shown. Algebraically $B_{5.2} = A_{5.2} + A_{5.4}A_{4.2} + A_{5.3}A_{3.2} + \cdots$.

The inverse input matrix element $B_{5.2}$ is essentially an accumulator of all these flows. That is, it indicates how much steel must be produced altogether—for all purposes—to satisfy one additional unit of demand of motor vehicles.

Since $B_{5.2}$ is an *input* matrix element, it expresses the fraction of all dollar *inputs* to motor vehicles that are attributable directly and indirectly to steel. Thus it immediately displays, among other things, the impact of a steel price increase on the costs of manufacturing motor vehicles. For instance, if steel accounts for 30% of the inputs (i.e., costs) of a car and steel prices rise by 10%, then costs of motor vehicle production will automatically increase by $0.3 \times 0.1 = 0.03$ or 3%. Presumably, this cost increase would be translated automatically into a corresponding price increase.

Now consider the inverse *output* matrix element $D_{5.2}$. It has exactly the same structure; that is, it has the same set of path products or chains:

$$D_{5.2} = C_{5.2} + C_{5.4}C_{4.2} + C_{5.3}C_{3.2} + \cdots$$

In this case, however, the interpretation is a little different. The direct output matrix element $C_{5.2}$ is the fraction of steel output that is sold directly to the motor vehicle industry, whereas $D_{5.2}$ is the fraction of all steel produced that goes directly or indirectly into making cars. This would not tell us how a steel price increase would affect the motor vehicle industry, but it will immediately show the impact of a change in demand for cars on the amount of steel required. Thus if motor vehicles utilize 0.2 (20%) of the output of the steel industry directly and indirectly, and motor vehicle output drops by 30%, then demand for steel will decline by $0.2 \times 0.3 = 0.06$ or

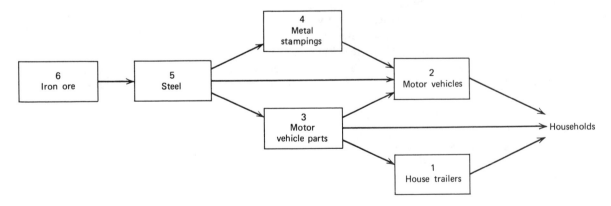

Figure 4.1. A simple set of chains.

6%. These numbers are fictitious, but they illustrate the uses of the two different types of inverse matrix elements.

Before concluding this section, it is worthwhile noting that the input–output model also provides relationships between the value added and the price, for each sector. Specifically, one can define the value added per unit output in the jth industry sector V_j as the price P_j received (per unit) less the total costs (per unit) of all purchased inputs, which can be expressed in terms of their prices, namely,

$$V_j = P_j - \sum_i A_{ij} P_i \qquad (4.13)$$

or in matrix notation

$$V = P[I - A] \qquad (4.13a)$$

These equations can, of course, be inverted, in the usual way,

$$P = V \cdot [I - A]^{-1} = VB \qquad (4.14)$$

The latter equations can be used to determine the price consequences of changes in industry costs due, for instance, to environmental regulation.

ADDENDUM TO SECTION 4.2

As noted in Section 4.1, it is possible to construct a nonsquare industry–commodity model that utilizes industry–commodity account statistics directly without the elaborate "balancing" procedure required by the conventional (square) I–O models, but at the expense of an additional postulate. Such a model has been built by Statistics Canada.[36] The current version utilizes 187 industrial sectors and 644 commodities. The basic definitions and equations are presented hereafter, maintaining a notation compatible with the one adopted previously. As before, the output of the ith industry (in dollars) will be denoted by capital letters X_i. Symbols representing commodity inputs and outputs, or commodity input

or output coefficients, will be denoted by lower case letters.

$$\text{Number of industries} = N$$

$$\text{Number of commodities} = n \qquad n > N$$

t_{ij} = dollar value of jth commodity *produced* by ith industry

$\tilde{t}_{ij}$ = dollar value of ith commodity *used* by jth industry

y_i = dollar value of ith commodity consumed by final demand

r_i = dollar value of the primary resource input to the ith industry

x_j = total dollar value of jth commodity

$$= \sum_{i=1}^{N} t_{ij} \qquad (4.15)$$

X_i = total dollar value of output of ith industry

$$= \sum_{j=1}^{n} t_{ij} \qquad (4.16)$$

Utilizing accounting balances (conservation of commodities)[†]

$$x_i = \sum_{j=1}^{N} \tilde{t}_{ij} + y_i \qquad (4.17)$$

$$X_j = \sum_{i=1}^{n} \tilde{t}_{ij} + r_j \qquad (4.18)$$

The assumption of unchanging technology (constant input requirements) defines the input coefficients:

$$\tilde{t}_{ij} = a_{ij} X_j \qquad (4.19)$$

whereas the assumption of unchanging market shares yields a counterpart equation defining the output coefficients:

$$t_{ij} = c_{ij} x_j \qquad (4.20)$$

[†]Some "commodities" in the Canadian system are actually services, for which no material conservation law is applicable.

Combining (4.16) and (4.20), we obtain

$$X_i = \sum_j^n c_{ij} x_j \qquad (4.21)$$

while (4.17) and (4.19) yield

$$x_i = \sum_j^N a_{ij} X_j + y_i \qquad (4.22)$$

Eliminating x_i from (4.22) gives

$$x_i = \sum_{j=1}^N \sum_{k=1}^n a_{ij} c_{jk} x_k + y_i \qquad (4.23)$$

In matrix notation, (4.21) and (4.22) become

$$X = cx \qquad (4.21a)$$

and

$$x = aX + y \qquad (4.22a)$$

$$= acx + y \qquad (4.23a)$$

These equations can be inverted, to give

$$x = (1 - ac)^{-1} y = by \qquad (4.24)$$

$$X = (1 - ca)^{-1} cy = dy \qquad (4.25)$$

where a and c are "input" matrices and b and d are "output" matrices bearing roughly the same relationships to each other as A, B, C, and D.

4.3 INPUT–OUTPUT AND NET ENERGY ANALYSIS

In Section 4.2 it was shown that the I–O formalism delineates sequences of transactions from raw materials through successive processing and manufacturing stages to final goods and services. As discussed earlier, some of these sequences represent "embodied flows" of materials that are eventually incorporated in final products, while others represent value added by "labor-saving" materials that are consumed en route, or by nonmaterial services.

It is clear that the total energy consumption of the economy can be allocated among the final goods and services produced, if the energy consumed by each sector is appropriately attributed to those sectors that buy from it, and so on to the last step in the sequence (final consumption). Thus each final product or economic service not only consumes energy (i.e., fuels and electricity) *directly* but also implicitly consumes energy that was actually purchased and used by suppliers and suppliers of suppliers at various stages. This can be thought of as *indirect* energy consumption by the final production question.[†]

The I–O formalism permits us to calculate the direct and indirect *dollar* contributions by all sectors to the outputs of any given final (or intermediate) sector. If A_{ij} stands for direct contribution, in dollars, of the product of the ith sector to the jth sector, we already have an expression that specifies the total of direct and indirect contributions of all other sectors, namely,

$$B_{ij} = \left[1 - A \right]_{ij}^{-1}$$

$$= \delta_{ij} + A_{ij} + \sum_k A_{ik} A_{kj} + \sum_{k,l} A_{ik} A_{kl} A_{lj} + \cdots$$

$$(4.11)$$

This expression is quite general. Obviously, it can easily be specialized to the case of final consumption.

To calculate the direct and indirect energy requirements associated with a dollar's worth of output of a given sector (say, the jth), it is merely necessary to attach appropriate energy consumption coefficients (in energy units per dollar) to each sector. Subject to the usual approximations of the input–output model itself, the procedure for calculating energy flows and indirect energy requirements by sector is absolutely straightforward.

The first type of energy coefficient that can be defined is as follows:

$$e_{kj} = \text{direct energy (of fuel type } k)$$

consumed by industry j

$$e_{kj} = \frac{E_{kj}}{X_{kj}} \qquad (4.26)$$

[†]Actually, the same logic applies also at intermediate stages. It is true for any product.

where E_{kj} is the total energy consumption, in energy units, by sector j from fuel sector k. The traditional fuel sectors are normally coal, petroleum products, natural gas, and electricity. ("Petroleum products" is sometimes further divided into gasoline and fuel oil.) In principle, E_{kj} must be given by the expression:

$$E_{kj} = \frac{C_{kj}X_k}{P_k} \qquad (4.27)$$

where C_{kj} is the output coefficient defined previously. Thus E_{kj} (in energy units) must be equal to the fractional output of the kth energy sector destined for sector j times the dollar output of the kth energy sector all divided by the unit price of that form of energy in *dollars per energy unit*. It happens that average fuel prices paid varies quite a lot from sector to sector due to differences in scale, geographical location, contracting terms, and so on. But this is merely a nuisance: It does not cause any difficulty in principle. In practice, E_{kj} is most easily derived directly from Census of Manufactures data.

The foregoing procedure focuses on fossil fuels and does not explicitly take into account energy obtained "free" from the environment, particularly solar energy. However, it is clear on reflection that solar energy inputs can be taken into account automatically by attaching energy coefficients to *all* organic materials of natural origin—not just fuels—such as food, cotton, wood products, and paper. This is generally not done for reasons of convenience, since such materials are not ordinarily thought of as part of the energy flow system. But there is no difficulty in principle. To trace energy flows beyond photosynthesis in green plants, for example, to the latent heat of water vapor in the atmosphere or solar radiation incident on the Earth, would also be possible, but hardly worthwhile. To do so we would need to add explicit sectors—for example, the water vapor cycle—that provide free goods and take up waste residuals according to physical laws, but neither purchase from, nor sell to, the economy per se. Since the activities of such a sector are not directly determined by the other sectors except in the very long run, there is little or no conceptual benefit in extending the boundaries of the input–output model so far at present.

Incidentally, I have been careful to refer to "energy units" without specifying any particular one, since there are some pitfalls to be noted and avoided. First, it must be pointed out that prices are usually given in terms of *volume* units (such as gallons, barrels, or cubic feet) or a *weight* unit (such as tons). To obtain prices in energy units, we need conversion factors for each fuel that give energy content per unit volume or unit weight, respectively.

But this does not take care of all the problems. It was pointed out in Chapter 3 that there are two quite different ways of defining energy "content." The more familiar (but potentially misleading) definition is "*total heat content*" (Section 3.1). This quantity is conserved according to the first law of thermodynamics in every process and can, therefore, be subject to accounting identities. The second possible definition of energy is "*available work content*" (Section 3.2), which is not a conserved quantity. However, from a physical standpoint, it is much more meaningful to keep track of available work consumption in each sector, than it is to keep track of total energy consumption. Of course, other definitions can be used, as long as it is done consistently.

It should be noted here that one of the reasons for confusion, in practice, between the concepts of total energy content and available work content, is the lack of any distinct conventional measure for the latter. In the absence of a better alternative, we might choose to measure available work in terms of "electrical energy equivalent" (e.g., kWh), since electricity is one of the two forms of energy that is, in principle, 100% available.[†]

[†]Kinetic energy of motion is the other. The availability of heat energy is a function of its absolute temperature T in relation to the temperature of the environment T_0. Heat energy is 100% available only when $T/T_0 \rightarrow \infty$. The availability of "chemical" energy can be calculated in terms of its maximum combustion temperature. The reason conventional space heating apparatus is so inefficient is that fuel is burned at a very high temperature ($\sim 2000°$F) in order to provide warm air at a temperature only a few degrees above the ambient temperature of the environment.

Clearly, many uses of energy can be classified according to the intrinsic quality of energy required. Thus space heating, cooking, and washing actually require very low quality energy (i.e., low temperature heat), whereas casting stainless steel or operating a "heat engine"—internal combustion or external combustion—requires very high temperatures, hence high quality energy. Many industrial uses are intermediate. Thus from a thermodynamic point of view, the most efficient use of hydroelectricity or fossil fuels, which are high quality sources, would be to "cascade" uses, using high quality waste heat, for example, from combustion gases, to provide energy for low grade requirements such as household uses. Notwithstanding the apparent logic of this approach, however, it must be emphasized that the state of maximum *thermodynamic* efficiency does not necessarily correspond with a state of maximum *economic* efficiency.

Available work content in relation to total energy content is, clearly, a measure of the "quality" of energy. As might be expected, the higher quality forms of energy (such as electricity) tend to command a much higher unit price than the lower quality forms. On the other hand, the correlation between quality and price is far from exact because of the existence of other factors in any given context. For instance, it may "pay" a developer to use electric baseboard heaters in an apartment complex because each unit is independent. Thus a breakdown in a central heating system could not simultaneously affect all the apartments. There are many situations where we might choose to use energy lavishly—hence inefficiently—in order to conserve capital or labor. (United States' agriculture is a cogent example of the latter).

The distinction between total energy content and available work content is theoretically quite important, though it does not have any practical consequences for an input–output model as long as *only* the traditional fossil fuel and electricity sectors are sources of energy flows to other sectors in the input–output model. The reason is that, for fossil fuels and electricity, both definitions of energy content are nearly equivalent. The available work content of a hydrocarbon fuel is within a few percent of being equal to its heat of combustion (in Btu), because it will burn at a very high temperature. Similarly, the available work content of electricity is precisely equal to its heat content, that is, the amount of heat (in Btu) that would be released by discharging the electric current through a resistive element, such as a toaster.

However, energy can be transferred from sector to sector by other means where the two definitions would not be equivalent at all. To take one possible example, if significant quantities of industrial or space-heating energy were supplied via purchased steam (rather than via purchased fuel or electricity), the total heat content of the inputs would differ dramatically from the available work content. Or, to take another example, a significant amount of energy is "embodied" in scrap materials. To the extent these materials are recycled, they constitute implicit energy inputs to the materials processing sectors (steel, aluminum, paper). Here, again, the two definitions of energy content are not at all equivalent. In any event, complete accounting of energy inputs to any sector should include energy embodied in scrap

or other nonfuel materials that are consumed in that sector.[†]

In any case, the energy of type k "embodied" in a dollar's worth of final demand for energy of type k can be defined as follows:

$$e_{ky} = \frac{E_{ky}}{Y_k} \tag{4.28}$$

$$E_{kj} = \frac{E_{kj}}{X_j} X_j = \frac{E_{kj}}{X_j} \sum_l (1 - A)_{jl}^{-1} Y_l$$

$$= \frac{E_{kj}}{X_j} \sum_l B_{jl} Y_l \tag{4.29}$$

where B_{jl} is the inverse input coefficient, defined previously. The total energy consumed by type is

$$E_k = \sum_j E_{kj} + E_{ky}$$

$$= \sum_{j,l} \frac{E_{kj}}{X_j} B_{jl} Y_l + \frac{E_{ky}}{Y_k} Y_k$$

$$= \sum_{j,l} e_{kj} B_{jl} Y_l + e_{ky} Y_k \tag{4.30}$$

A new matrix (the "energy matrix") can be defined by

$$E_{kl} = f_{kl} Y_l \tag{4.31}$$

where
$$f_{kl} = \sum_j e_{kj} B_{jl} \qquad l \neq k$$

$$f_{kk} = \sum_j e_{kj} B_{jk} + e_{ky} \qquad l = k \tag{4.32}$$

It can be seen that the e_{kl} are direct energy coefficients where the f_{kl} are the corresponding indirect coefficients. (Note that the indirect energy coefficients are obtained by matrix multiplying the indirect dollar coefficients by the corresponding direct energy coefficients.)

Actual numerical values for the direct energy coefficients for a 35 sector input–output table (U.S.)

[†]This is particularly important in industries such as lumber, paper, and food processing that obtain part of their energy requirements from waste nonfuel organic materials.

have been calculated by Reardon[37] for 1947, 1958, and 1963 for the purpose of assessing time trends in the use of energy by different sectors.[†] Recently, Herendeen has calculated complete sets of direct and indirect energy coefficients for 367 sectors (OBE) for the years 1963 and 1967.[39] These coefficients have been used by Herendeen and Bullard to assess the relative energy intensity of a variety of goods and services.[40]

The Herendeen coefficients have also been applied to a detailed analysis of the direct and indirect use of energy in automotive transportation by Ayres, Betlach, and Decker.[41] In the latter study, the coefficients were aggregated to a smaller (185-sector) I–O table[‡] and extrapolated to 1971. Both the (then) current configurations and mix of automobiles, and a number of alternatives technologies were considered and compared.

As can be readily appreciated, the comparison of hypothetical technological alternatives introduces a new set of complexities to the analysis. Each alternative case implies a different set of dollar I–O coefficients, especially among those sectors that sell directly or indirectly to the motor vehicle manufacturing sector. For example, a decrease in the size of vehicles involves a reduction in the weight of steel that is purchased, but not necessarily a corresponding reduction in the price.[§] To substitute aluminum and/or plastic for steel involves an increase in purchases from the one and a decrease from the other. Again, the shifts are not proportional, either in money or in dollars, since a given weight of aluminum replaces a larger (but variable) weight in cast iron or steel, depending on the components affected.

Fortunately, the number of sectors for which the A_{ij} must be recomputed for each technology is only of the order of a dozen or so. Obviously each such alternative case involves a corresponding recalculation of the indirect dollar coefficient. Here the series approximation method discussed in Section 4.2 [Equations (4.9) to (4.12)] is vital, since it enables us to compute the indicated changes in the inverse

element B_{ij} directly from the finite number of specified changes in the A_{ij}. (It is *not* necessary to invert the matrix each time.)

Figures 4.2 to 4.5 display some representative results of the study. Figure 4.2 shows the overall energy relationships of the industry for the "base" case (1972), expressed in Btu per passenger mile. Figure 4.3 shows a breakdown of the manufacturing energy among various subsidiary chains of suppliers. Figure 4.4 gives a corresponding breakdown of the energy inputs to the gasoline-refining and auto-repair sectors. A selected sample of technological alternatives are compared in Figure 4.5.

From the overall economic viewpoint, the general scheme described in the foregoing paragraphs provides adequate detail. However, it must be noted that only *six* energy sectors are explicitly considered in the widely used INFORUM model,[26, 42] namely,

coal mining (14)
crude oil and natural gas (15)
petroleum refining and related products (69)
heating oil (70)
electric utilities (160)
natural gas (161)

Of course, the input–output table indicates dollar transactions among these sectors, and between them and the nonenergy sectors of the economy. Herendeen et al. have determined the Btu equivalents of these flows. But for many analytic purposes, greater detail is needed. To achieve this, the foregoing energy sectors can be detached from the input–output model and replaced by a physical network model of energy fuel extraction, processing, and conversion, by fuel type and "end-use." Brookhaven National Laboratories has developed such a model for the Energy Research and Development Administration (ERDA), called the Energy System Network Simulator (ESNS).[43] The Brookhaven model has been adapted for use in conjunction with a modified (110 sector) version of the Herendeen model, which is, in turn, "driven" by a highly aggregated (13 sector) input–output model developed by Hudson and Jorgenson[44, 45] to determine the equilibrium of the economy under conditions of changing energy prices. A slightly more elaborate, but closely related version of this scheme, has been developed for the U.S. Environmental Protection Agency (EPA) by Narkus-Kramer et al.[46] for use in conjunction with the INFORUM model to

---

[†]Reardon's work was anticipated to some extent by Strout.[38]
[‡]The Interindustry Forecasting Model of the University of Maryland (INFORUM) developed by Clopper Almon, Jr.[42, 26] A list of the INFORUM sectors together with corresponding sectors of the BLS model and their SIC counterparts is given in the appendix to this chapter. All sectors specifically identified by number refer to the INFORUM numbering system.
[§]In fact, the "standard" assumption of I–O is clearly invalid here.

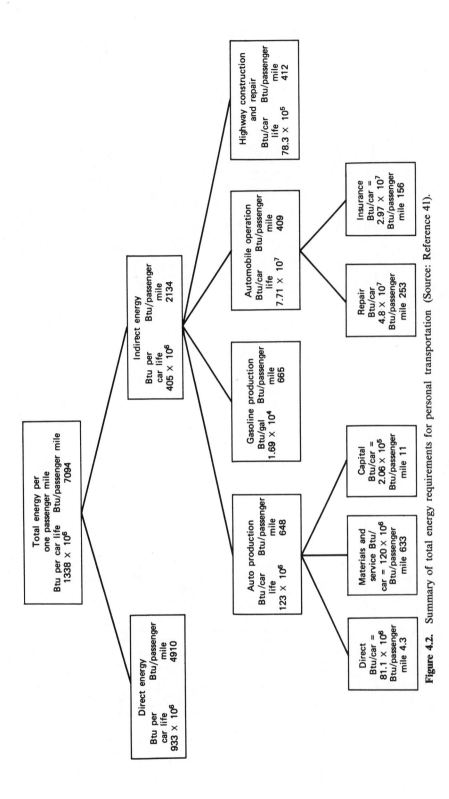

Figure 4.2. Summary of total energy requirements for personal transportation (Source: Reference 41).

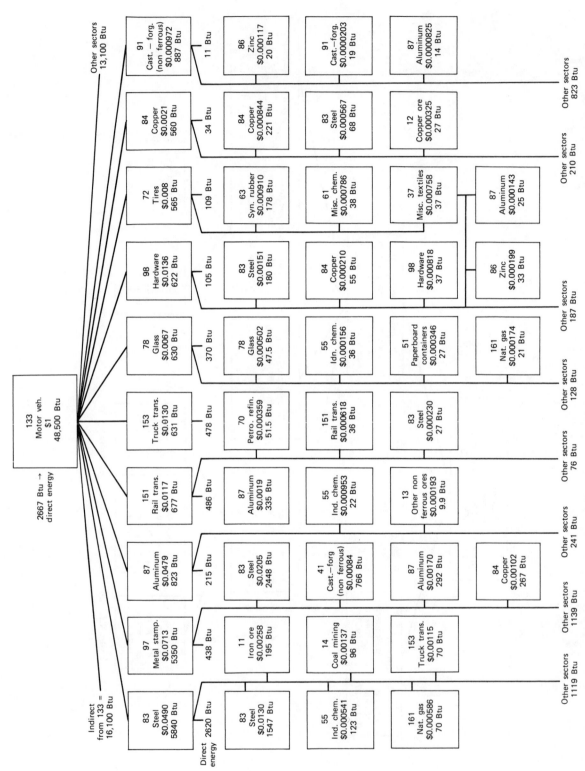

Figure 4.3 Major direct and indirect commodity inputs to auto manufacturing and their corresponding energy requirements (Source: Reference 41).

108

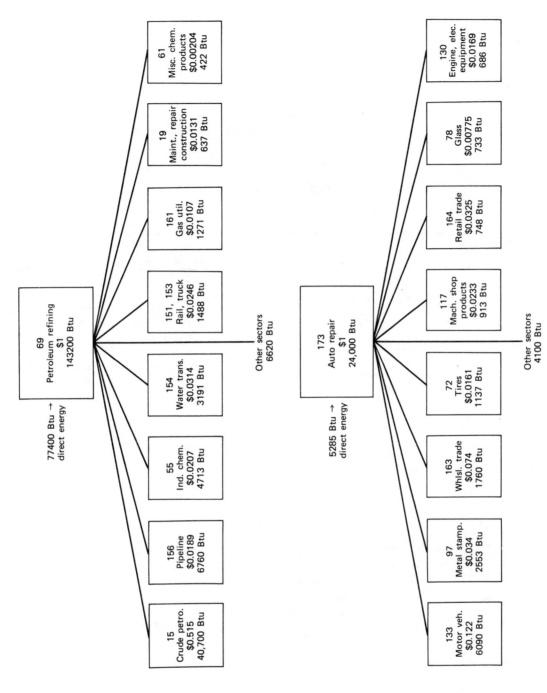

Figure 4.4. Major energy contributory to automobile fuel production and repair (from Almon 1971 and Herendeen 1963; Source: Reference 41).

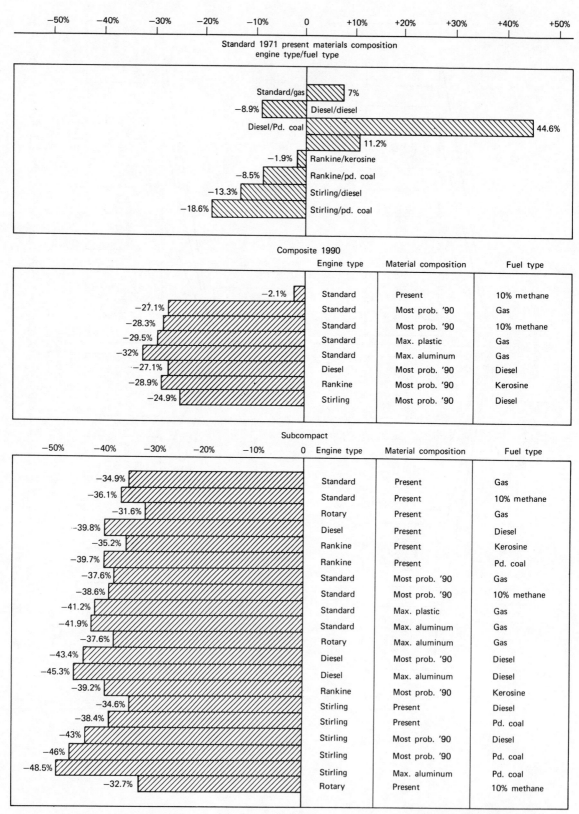

Figure 4.5. Percentage difference from base case in total lifetime Btus of selected substitutions (Source: Reference 41).

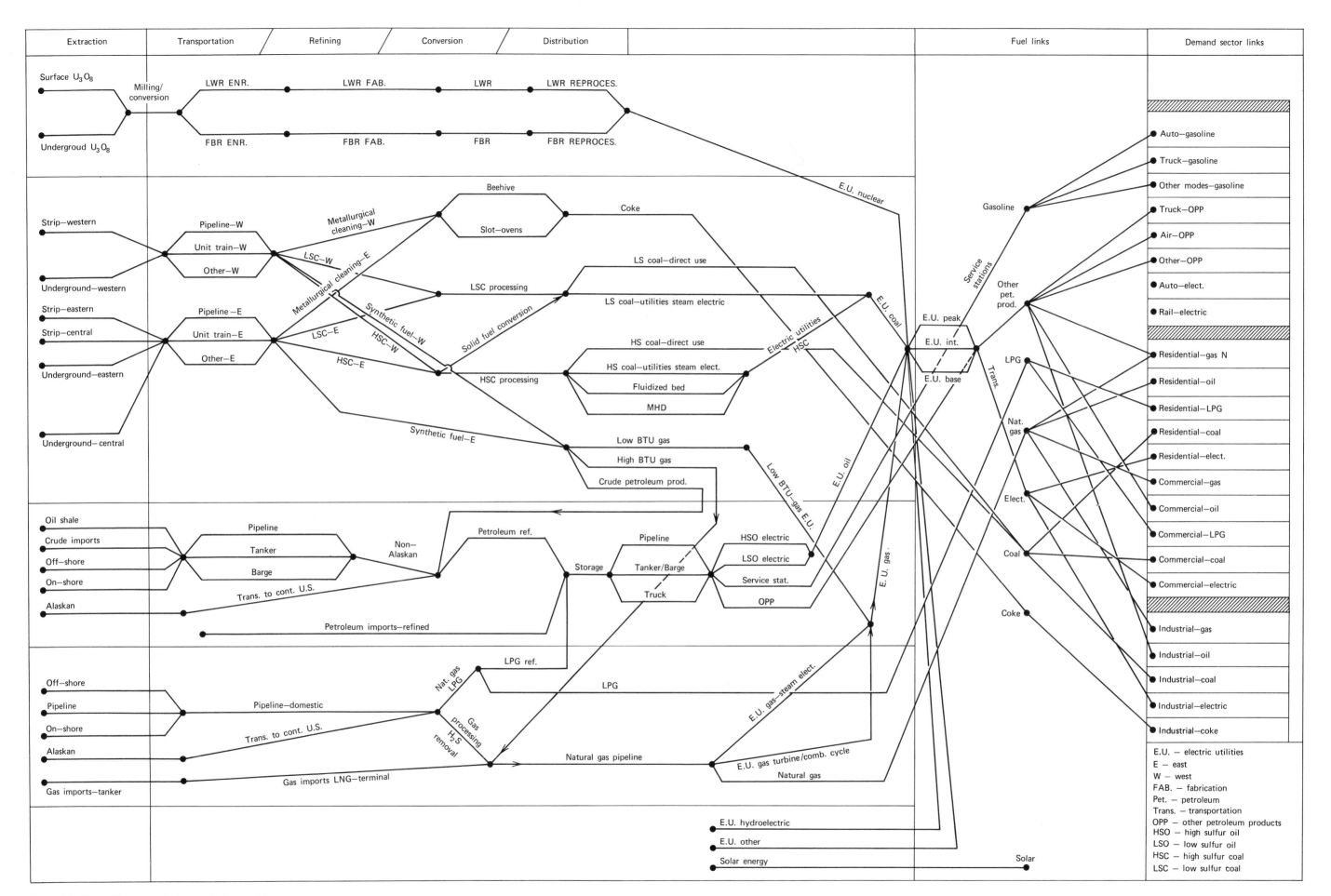

Figure 4.6. Energy flow network (Source: Reference 50).

assess both energy and environmental implications of alternative national policies and economic growth scenarios. The latter scheme is shown graphically in Figure 4.6.

4.4 MATERIALS FLOW ANALYSIS USING INPUT–OUTPUT MODELS

As indicated in Section 4.1, the industry–commodity accounts contain much of the structural information that would be needed to track materials flows (and transformations) through successive stages of the economy. However, quantitative physical flows can only be estimated from dollar volumes if reliable average price data for each commodity category is available. Even in a fairly disaggregated case, such as the Canadian Industry–Commodity Table (187 industries and 644 commodity sectors), a "commodity" is generally a composite of many commercial grades or varieties.

To complicate matters, published wholesale prices are quoted in a variety of different ways, sometimes including and sometimes net of shipping costs, demurrage, insurance, excise taxes, brokerage charges, and the like. "Spot" market prices differ significantly from long-term contract prices for some raw materials (e.g., coal, lumber, copper, etc.). For these and other reasons, no comprehensive standardized and satisfactory file of commodity prices is available.

To predict the pattern of physical materials use, therefore, the approach that has been followed in most studies to date has been to construct empirical equations relating demand for specific materials in physical units (e.g., tons) to aggregate measures of economic activity. In most widely known studies (by Resources for the Future, Inc.[47, 48] and the Bureau of Mines[49]), use of materials was not specifically related to industry sectors at all, but to aggregate demand categories such as housing, automobiles, consumer durables, and producer durables. However, in more recent United States studies,[50, 51] efforts have been made to tie physical flows of materials more closely to the activities of industrial sectors, which in turn are driven by aggregate demand.

The method adopted in Reference 50 makes explicit use of the fact (discussed in Section 4.2) that the inverse elements of a matrix of I–O coefficients can be expressed as the sum of a set of "path

products," namely,

$$B_{ij} = \delta_{ij} + A_{ij} + \sum_k A_{ik}A_{kj} + \sum_{k,l} A_{ik}A_{kl}A_{lj} + \cdots$$

$$(4.11)$$

where the inverse element B_{ij} represents the total of direct and indirect inputs to sector j originating from sector i. In this case, we suppose that sector i is a "basic material" sector (e.g., steel), whereas sector j is any final product or service sector (e.g., automobiles) that sells to final demand. It can be assumed that materials flow M_{ij} in physical units from sector i to sector j is proportional to B_{ij} (subject to certain adjustments, discussed later)

$$M_{ij} = k_{ij}B_{ij(\text{adj})} \qquad (4.33)$$

where k_{ij}^{-1} is clearly a "unit price" of some sort.

As remarked previously, the path products (or chains) of transactions on the right-hand side of Equation (4.33) represent indirect contributions of the output of sector i to sector j. Some of these chains reflect flows of embodied materials, whereas other flows reflect labor-saving materials that are consumed (and discarded) en route, or nonmaterial services between sectors. If one is interested primarily in the embodied flows, it is generally possible to pick these out by inspection of the path products.

For example, suppose sector i refers to steel (INFORUM Sector 83) and sector j refers to "motor vehicle manufacturing" (INFORUM Sector 133), the sector that produces automobiles. Obviously, some iron and steel is sold directly by sector 83 to sector 133. But a significant amount passes first through other sectors, such as stampings, bolts, springs, and bearings. The 20 most important transaction chains linking these two sectors are displayed in Table 4.4.

In examining the 20 largest path products in the inverse element $B_{83.133}$, we can immediately identify some peculiarities that warrant closer examination. These are of two kinds:

1. Loops (chains 11, 12, 13, 14, and 19).
2. Chains with spurious links of some kind (involving products not actually incorporated in automobiles or used by the auto industry).

Of the 20 largest chains (in terms of magnitude), no

Table 4.4 Twenty Most Important Transaction Chains from Steel (83) to Motor Vehicles (133)

Chain	Values	Cumulative Sum	Sectors[a]		
1	0.066519	0.066519	83	133	
2	0.009994	0.076513	83	97	133
3	0.001707	0.078220	83	96	133
4	0.001250	0.079470	83	99	133
5	0.001151	0.080621	83	111	133
6	0.001026	0.081647	83	117	133
7	0.000621	0.082268	83	101	133
8	0.000392	0.082660	83	102	133
9	0.000261	0.082921	83	130	133
10	0.000218	0.083139	83	124	133
11	0.000153	0.083292	83	101	83	133
12	0.000118	0.083410	83	96	83	133
13	0.000112	0.083522	83	117	83	133
14	0.000103	0.083625	83	11	83	133
15	0.000093	0.083718	83	72	133
16	0.000086	0.083804	83	142	133
17	0.000062	0.083866	83	96	97	133
18	0.000060	0.083926	83	92	68	133
19	0.000040	0.083966	83	100	83	133
20	0.000038	0.084004	83	101	97	133

Sectors

11	Iron ore	100	Valves, pipe fittings, and pipe
68	Paint	101	Steel springs
72	Tires and tubes	102	Engines and turbines
83	Steel	111	Ball and roller bearings
92	Metal cans	117	Machine shop products
96	Screws and bolts	124	Electric lighting and wiring
97	Stampings	130	Engine electric equipment
99	Upholstery springs	133	Motor vehicles
		142	Mechanical measuring devices

[a]In this table and throughout the subsequent discussion, the sectors are numbered in accordance with definitions used in the University of Maryland Interindustry Forecasting Model (IN-FORUM) developed by Clopper Almon, Jr., et al. The list of INFORUM sectors and corresponding four-digit SIC sectors is given in the appendix to this chapter.

Source: Reference 52.

less than five involve sector 83 (steel) as an *intermediate* as well as a starting point. Chain 11 (steel → springs → steel → autos) probably refers to forged leaf-springs, since iron and steel forging establishments are counted as part of the basic iron and steel industry. This loop would be eliminated if forging were segregated (like stampings) from the rest of the iron and steel sector. Loops 12, 13, and 19 also probably involve forging or heat treatment of bolts, machine shop products, and valves, respectively. However, chain 14, involving the iron ore sector, is difficult to interpret and probably should be eliminated.

A good example of a spurious linkage is found in chain 18, which involves the sale of steel to metal cans (92) to the paint industry (68) to the motor vehicle industry. The problem is that the metal cans are not incorporated in the next product (paint), nor does the auto industry buy paint in metal cans. The cans are evidently containers for paint sold to final consumers, where they are discarded as solid waste. Another spurious link is found in chain 8 (steel → engines and turbines → motor vehicles). The problem here is that sector 102 includes diesel engines and aircraft gas turbines, but not gasoline engines. The engines bought by the motor vehicle industry from sector 102 go into heavy trucks and buses, which are sold to the commercial transportation industry, not into private cars, which are sold to consumers.

Evidently, by examining the chains individually one can pick out and eliminate those that are inappropriate for some reason. The procedure is necessarily ad hoc. It is desirable, however, to tighten the causal connection between final demand and material consumption, by eliminating as many spurious links as possible. Often, it is the links involving service sectors (e.g., insurance) that are discarded. Of course, service sectors do purchase material products (especially paper) in proportion to the demand for their services. But, while they do sell to other sectors, and not entirely to final demand, it is helpful to assume the contrary from the standpoint of deriving empirical equations for materials flow.

In order to utilize Equation (4.33), one must also estimate the multiplier k_{ij} for some convenient base-year, and (if possible) assess its rate of change over time. The first of these can only be accomplished by actually determining the allocation of materials among final products—as well as among labor-saving intermediate uses (such as acids solvents and industrial detergents)—and constructing a materials–product matrix. Such a matrix for 1972 is illustrated in Table 4.5.[50] While the available data is incomplete and unreliable in various ways, we can take advantage, as always, of the (first law) accounting identities that state, in effect, that all physical outputs of the materials producing sectors—allowing for imports and exports—must be either embodied in final products or discarded as intermediate wastes.[†]

[†]This disregards year-to-year changes in stockpiles and inventories.

Table 4.5 1972 (Base Year) Materials–Product Flow Matrix (in units of 10^3 tons)

Basic material	Code	Newspaper (01)	Writing/ printing (02)	Tissue and other paper (03)	Packaging/ container (04)	Construction (05)	Other paperboard products (06)	Household durables (07)	Electrical/ electronics (08)	Furniture (09)	Other consumer goods (10)	Auto and light trucks (11)	Other transportation (12)	Other plastics (13)	Other glass products (14)	Other textile products (15)	Machinery (16)	Other metal products (17)	Batteries (18)	Apparel (19)	Tires (20)
Aluminum	01				539	1,461		514	701			407	610				342	377			
Copper	24					432			1,252			159	68				300	198			
Ferrous metals	31				6,147	21972	1,964	2,647		512	20,997	4,379					33,791				
Glass	35				11,874	1,642			1,206		1361	1,361	3		1,244	9,800					
Lead	44				33	88		60	128			22				288		99	727		
Paper	58	10,150	9,890	4,200	5,770																
Plastics	62				2,604	2,754		386	736	502	964	534	290	2,705							
Rubber	65								33	109	404	87					218			240	2,071
Textiles	72										681	59				2349				2,084	338
Wood	79				5,741	47,001				3,298	1,092						697				
Zinc	80					519			176			344					137	179			
Paperboard	81				19,923		1872														

113

Having appropriately adjusted the terms in Equation (4.33) and fixed the actual materials allocation among product classes for a base year, the forecast itself is driven (as usual) by the assumed trajectory of macro-economic variables and aggregate consumption.

The approximate methodology described here is quite suitable for analyzing *gross* flows of materials in forecasting future solid waste generation, for instance. However, some of the most significant materials in the economy—such as specific chemicals or minor metals—cannot be correlated satisfactorily to intersectoral transactions at the level of detail shown in Table 4.5 or 4.6.

It is true that the path product approach offers a considerably greater degree of refinement than most critics of I–O have assumed. As an example, consider the demand for silver—a minor part of the primary "nonferrous metals nec" (sector #88). Obviously, demand for silver is not closely related to total output of sector #88. However, the two major users of silver are in photographic film (sector #145) and "jewelry and silverware" (sector #147). It is quite reasonable to postulate that the major demand for silver will be driven primarily by several specific chains (88-147) and (88-55-145) or (88-55-61-145). Note that silver for the manufacture of film would not be purchased directly from sector 88, but via the industrial chemical industry (55) or the chemical products industry (61).

To take one more example, consider the metal tungsten, which is used mainly for tungsten carbide or "high speed alloys" in cutting tools (although it has a variety of other miscellaneous small uses). This metal is classed as "ferrous," but the demand for tungsten is not necessarily closely related to the output of basic iron and steel (sector 83). However, the output of "metal cutting machine tools" (sector 106) is much more relevant to the use of tungsten. And we can even distinguish the sale of replaceable tungsten carbide drillbits by sector 106 to other sectors, such as "machine shop products" (sector 117) or "engines and turbines" (sector 102), from the sale of complete machine tools, since the former is a short-lived maintenance item, whereas the latter is a capital investment and thus treated as an element of final demand. Thus the demand for tungsten would be driven primarily by a few path products such as (83-106-102), (83-106117, or (83-106-133).

Although a surprisingly large amount of information can be extracted from the I–O table, with the

help of the type of path product analysis already illustrated, there are obviously limits to what can be done with those tools. The major problem, without question, is the chemical industry, where literally thousands of different chemicals are produced, many of them consumed within the sector itself. Even where chemicals are sold extensively outside the chemical industry, for example, to pharmaceuticals (sector 66) or paints and pigments (sector 68), the path products reveal nothing about any individual chemical.

The only satisfactory solution in this case is to disaggregate the sector to the level of individual chemicals. A physical input–output model covering 395 chemicals of the industrial chemical industry *per se* has been developed.[53] Of course, this is a formidable enterprise, as a consequence of the enormous amount of data that must be compiled, much of it from trade publications or other nongovernment sources.

The basic equations of the chemical input–output model are essentially the same as those for the conventional (dollar flow) model, except that all sales outside the chemical industry are treated like "final demand" and all inputs and outputs are given in physical units (e.g., tons). For convenience, I use Greek letters to distinguish the variables in the chemical input–output model (CIOM) from their counterparts in the industry–industry model and the industry–commodity model.

Let

$$\sum_j \tau_{ij} + \xi_i = \chi_i \qquad (4.34)$$

Where τ_{ij} are tonnage sales of chemical product i to producers of other chemical products j (summed over all chemical products), ξ_i represents sales of i to all end-users outside the chemical industry,[†] and χ_i represents total output of the ith chemical. This is a perfectly general equation.

As before, the standard input–output hypothesis would be

$$\frac{\tau_{ij}}{\chi_j} = \alpha_{ij} \qquad (4.35)$$

where α_{ij} represents the fraction by weight of all inputs to chemical product j that come from chemi-

[†]Actually, the CIOM distinguishes sales to other intermediate sectors and sales to final consumers (see Table 4.7).

Table 4.6 Representative Chemical Output Coefficients

Chemical	Uses within the chemical industry												Use by other sectors										
	VCM	Trichlorethylene	Perchlorethylene	Vinyl acetate	Acrylonitrile	Acrylic acid esters	Cellulose acetate	Acetic esters	Chloroacetic acid	ABS	Fumaric acid	Hydroquinone	Pesticides and agr (60)	Misc. Organics (61)	Plastic materials (62)	Synthetic rubber (63)	Noncellulosic fibers (65)	Drugs (66)	Petroleum refining (69)	Rubber products (73)	Other sectors	Final domestic demand	Exports
EDC	0.77	0.03	0.03																0.03		0.06	0.00	0.08
Acetylene	0.26			0.21	0.08	0.09										0.27					0.09		
Acetic acid				0.31			0.44	0.14	0.03												0.08		
Acrylonitrile										0.17						0.06	0.61				0.16		
Maleic anhydride											0.16		0.10	0.16	0.14		0.40				0.04		
Aniline												0.09		0.15	0.06			0.06		0.60			0.04

115

cal product i. Substituting (4.36) in (4.35), we obtain

$$\sum_j \alpha_{ij}\chi_j + \xi_i = \chi_i$$

or

$$\xi_i = \chi_i - \sum_j \alpha_{ij}\chi_j \qquad (4.36)$$

which can also be expressed in matrix form as

$$\xi = (I - \alpha)\chi \qquad (4.36a)$$

The inverse relationship is

$$\chi = (I - \alpha)^{-1}\xi = \beta\xi \qquad (4.37)$$

An element of the inverse matrix, say β_{ij}, represents the total of all direct and indirect needs for chemical i to produce a unit quantity of chemical j. In the chemical industry, indirect pathways are not especially numerous,[†] but important examples exist. The chemical ethylene dichloride (EDC), for instance, can be converted by pyrolysis directly into vinyl chloride monomer (VCM), but the pyrolysis process also generates hydrochloric acid (HCl) as a by-product, which, in turn, can be reacted with acetylene to yield vinyl chloride by another route. The direct matrix element $\alpha_{EDC, VCM}$ reflects the production of VCM by means of the EDC pyrolysis process. The indirect element $\beta\{EDC \rightarrow VCM\}$ also includes the secondary route by way of hydrochloric acid:

$$\beta_{EDC, VCM} = \alpha_{EDC, VCM} + \alpha_{EDC, HCl} \cdot \alpha_{HCl, VCM}$$

It is important to note that these relationships can be derived from a technical knowledge of the structure of the chemical industry. Moreover, if a product is produced by only one process (from a unique set of raw material intermediates) the numerical values of the α-coefficients can be inferred from a knowledge of the chemical reaction involved and the atomic weights of the constituents. For instance, exactly 245 units of sulfuric acid (H_2SO_4) and 195 units of fluorspar (CaF_2) are required to yield 100 units of hydrofluoric acid (HF) plus 340 units of waste calcium sulfate ($CaSO_4$). These ratios can be determined from the reaction chemistry and the atomic weights of the elements involved.

[†]The chemical input–output table is rather "sparse" in the strict mathematical sense.

Output coefficients can also be defined, as in the previous case,

$$\frac{\tau_{ij}}{\chi_i} = \gamma_{ij} \qquad (4.38)$$

whence

$$\xi_i = \chi_i - \sum_j \gamma_{ij}\chi_i \qquad (4.39)$$

or

$$\xi = \chi(I - \gamma) \qquad (4.39a)$$

Solving for χ, we obtain

$$\chi = \xi(I - \gamma)^{-1} = \xi\delta \qquad (4.40a)$$

The matrix of output coefficients γ and the inverse output matrix δ have corresponding interpretations. As noted previously, either form of matrix can be derived from the other. For instance, if the published data most readily available for each chemical are the total production and the breakdown of uses, both for manufacturing other chemicals and for other purposes, then we can fill out a table of direct output coefficients γ_{ij}, as illustrated in Table 4.6. The corresponding input coefficients can be computed by equating (4.37a) and (4.40a) and formally solving for α

$$\alpha_{ij} = \delta_{ij} - \sum_{k, l} \frac{\chi_{ik}}{\chi_{lj}}(\delta_{kl} - \gamma_{kl}) \qquad (4.41)$$

or, in matrix form

$$\alpha = I - \chi(I - \gamma)\chi^{-1} \qquad (4.41a)$$

Of course, these calculated input coefficients should agree with what one would compute from a detailed analysis of the process (or processes) used to manufacture each chemical product. Similarly, one could reverse the procedure and derive the output coefficients from published production statistics and chemical process data from engineering sources. Either type of data, together with materials balance considerations, can be used to correct errors or omissions in the other.

As an example, consider the production of vinyl chloride monomer (VCM) from ethylene dichloride

(EDC), previously discussed. The following example illustrates the use of materials balance conditions to uncover errors.

- According to Reference 54, United States production of EDC, VCM, and PVC in 1972 was as follows:

Ethylene dichloride ($C_2H_4Cl_2$)   (EDC) 7809×10^6 lb
Vinyl chloride Monomer (C_2H_3Cl)   (VCM) 5089×10^6 lb
Polyvinyl chloride and copolymers   (PVC) 4322×10^6 lb

- Utilization of EDC for that year was estimated by trade sources[55] as follows:

VCM   77%
Trichlorethylene   3%
Perchloroethylene   3%
Petroleum Refining   3%
Exports   8%
Other   6%

- Utilization of VCM was estimated to be:

PVC and Copolymer   90%
Exports   10%

- A materials balance can be established for the amount of VCM that should have been produced from the available EDC, based on the reaction chemistry:

$$EDC(C_2H_4Cl_2) \rightarrow VCM(C_2H_3Cl)$$
$$+ HCl + Wastes + unreacted\ EDC$$

The molecular weights of EDC, VCM, and HCl are 99, 62.5 and 32.5, respectively. Assuming EDC is converted to VCM with 95% process efficiency, it requires about $1/0.95 \times 99 = 104$ (molecular) units of EDC to yield 62.5 units of VCM. Thus 1.67×10^6 lb of EDC will suffice to produce 1.00×10^6 lb of VCM. Therefore, from the stated EDC production level of 7809×10^6 lb, the production of VCM from EDC should have been

$$\frac{7808.9 \times 10^6(0.77)}{1.67} = 3600.5 \times 10^6\ lb$$

- An estimate can now be made for total VCM output, assuming 92% of VCM is made from EDC. (The remainder of VCM production employs acetylene as its principle intermediate material.)

$$\frac{3600.5 \times 10^6}{0.92} = 3913.6 \times 10^6\ lb$$

This is considerably less than the stated 5088.5×10^6 lb VCM production. Since the materials balance condition is reasonably precise, the error is likely to be either in the reported total, production figures for EDC or VCM, or in the reported percentage of VCM attributable to EDC.

- Compatibility between the VCM and PVC production figures can also be checked. Assuming that 5% material losses occur during the polymerization step and that 90% of domestic VCM production goes to domestic PVC and copolymer production, total PVC output should be

$$\frac{5088.5 \times 10^6(0.9)}{1.05} = 4361.6 \times 10^6\ lb$$

This is in relatively good agreement with the published figure for PVC production of 4322.0×10^6 lb. Since the PVC and VCM figures are consistent with each other, the error would seem to be in the EDC figures.

- Taking the VCM and PVC figures as accurate (as well as the fractional attribution of VCM to EDC), a revised estimate for EDC production can now be obtained:

$$7809 \times 10^6 \frac{5089}{3914} = 10{,}150 \times 10^6\ lb$$

This figure is 30% higher than the published value for EDC prodution. The adjustment is believed to be accurate on the basis of the foregoing material balance analysis. It is to be noted that only 18% of the announced EDC production enters the merchant market. The remainder is consumed at the plant site, which increases the likelihood of reporting errors.

It is not strictly necessary to assume that either the input coefficients (α_{ij}) or the output coefficients (β_{ij}) are always constant over time. However, the input coefficients are definitely linked to the process technology in use in the industry and only change as processes substitute for each other. For instance, processes based on ethylene are steadily displacing processes utilizing acetylene for the production of VCM. Monitoring chemical output statistics from year to year and using the relationships already defined would permit us to infer the pace of ongoing technological changes within the chemical industry.

4.5 EXTENSION OF INPUT–OUTPUT MODELS TO INCLUDE ENVIRONMENTAL LINKS

Perhaps the first authors to systematically approach the resource/environment interface from an economic perspective were Herfindahl and Kneese.[56] Later Kneese and the author proposed a general equilibrium materials balance model in which economic activities were linked to the environment by residuals generation and waste management.[57] Actually, Cumberland[58] seems to have been the first to suggest the extension of conventional I–O models to incorporate environmental factors in analyzing alternate regional development strategies. Cumberland has continued to contribute in this area, particularly emphasizing the use of I–O models to generate gross residuals (on a regional or national basis) by means of a matrix of residuals coefficients, and employing benefit–cost analysis to select optimal pollution abatement strategies.[59–61]

An elegant theoretical formulation of the extended I–O model to include both residuals generation and abatement activities in a single framework was published by Wassily Leontief in 1970.[62] A partial application to the air pollution problem was presented by Leontief and Ford the following year.[63] Richard Stone provided a useful tutorial paper that emphasized the evaluation of abatement costs and the choice of optimum abatement levels.[64]

The first use of an I–O model for purposes of forecasting comprehensive future economic and environmental implications of several alternative national policies over an extended time frame was carried out for the Commission on Population Growth and America's Future by Ridker and others at Resources for the Future, Inc., in collaboration with the present author and several colleagues.[65, 66] Since that, time there have been several other similar projects including the United States Environmental Protection Agency's SEAS[†] program[67, 68] and models by Victor in the United Kingdom[69] and Thoss in Germany.[70] The UN World Model developed by Leontief et al. is an important step forward in several respects.[71, 72]

An extension of the I–O approach to incorporate ecological relationships per se was attempted in 1967

by Isard and his colleagues.[73] Subsequently, there have been several ambitious projects along these lines.[74–76] But, despite major efforts, it appears that the necessary theoretical understanding and data for a satisfactory, comprehensive, "macro-economic-ecologic" model do not yet exist. (This does not apply, of course, to small scale ecological models, of which there are now many interesting and useful ones.)

All extensions of the input–output model to include residuals flows utilize roughly the same technique, with minor variations. I describe a scheme similar to that of Leontief,[62] but with one modification and a further elaboration to deal explicitly with raw materials *extracted* from the environment as well as waste materials *returned* thereto.

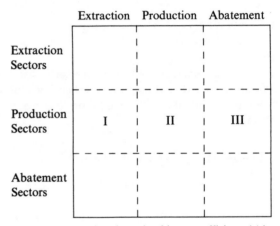

Figure 4.7. Partitioned matrix of input coefficients (A_{ij}).

First, the list of industrial sectors is extended to include abatement activities. The (still square) input–output matrix is then partitioned into three column groupings labeled I, II, and III, as shown in Figure 4.7. The A_{ij} coefficients in the "extraction" column (I) are, of course, already in the conventional I–O matrix. They represent the fractional inputs (in dollars) per unit output (in dollars) of each sector. The coefficients in the "abatement column" (III) have a similar interpretation; the output of each sector is, of course, a *service*, comparable to any other service in the economy.[†] The revenues accruing to each abatement sector would simply be

[†]SEAS Strategic Environmental Assessment System. This is actually based on the INFORUM model,[46] modified and disaggregated in various ways. Supplementary modules have been built to estimate waste residuals, resource prices, supply constraints, regional allocations, land-use impacts and the like.

[†]For purposes of the model, the abatement sectors need not be *institutionally* separate; in practice it is only necessary that the inputs and costs associated with abatement be separately accounted for.

the *costs* of abatement (to whatever level is actually achieved).

It is not necessary to define a "unit" of abatement at this stage, although one would have to do so in order to compute price effects (as is discussed later). In this respect, the scheme being described differs from that of Leontief, who defines the coefficients in column(s) III in terms of dollar inputs (i.e., costs) per physical unit (e.g., gram or kilogram) of pollutant removed.

Note that in the 185 sector INFORUM system there are 12 "extractive" industries including agriculture, forestry and fishing, and mining and drilling, but excluding meat, milk, and egg production. The most likely classification of abatement sectors would be by discharge media, namely, air pollution treatment, water pollution treatment, and solid waste management, including recovery of secondary materials from consumption wastes. Thus there might be three abatement sectors. (Leontief apparently envisioned a sector for each pollutant; this is possible but not necessary.)

The next step is to define a resource input matrix R and a waste output matrix W (each of which is also partitioned into three smaller segments), linked to the capital I–O matrix as shown in Figure 4.8

The Resource Input matrix R (IV, V, and VI) would have one row for each raw material or other resource distinguished in the system, and one column for each industry sector in the central I–O matrix. (The latter might be a nonsquare industry–commodity matrix, in which case there would be one column for each commodity.) The matrix element R_{ij} would give the resource input (in physical units) per dollar of output of the sector (or commodity). Presumably the list of materials (rows) would include air and fresh water, a number of cultivated crops (or crop types), fish caught, trees cut, minerals quarried or mined (by type of ore), fossil fuels, and so on. Land use, by type, might also be included. Most of these resources are collected by the extraction sectors, of course, so that the majority of the nonzero entries would be in the upper left-hand corner (IV). But there would be entries for other resource flows,

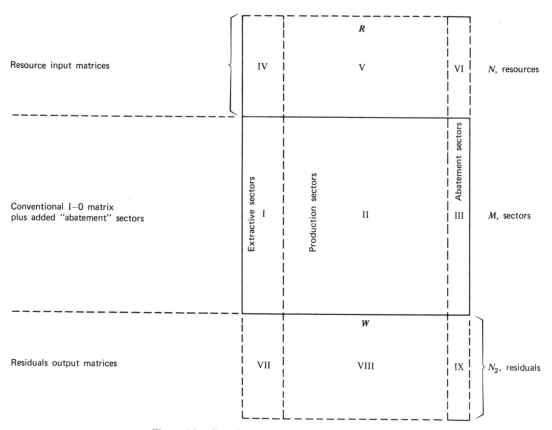

Figure 4.8. Complete system of partitioned matrices.

including the use of so-called free goods, such as air, for combustion or water for dilution, washing, and the like. Recycled scrap or recovered waste materials from the abatement sectors also appear as inputs in the appropriate columns. Each entry would be in physical units of input (e.g., kilogram) per dollar of output of the sector.

A variant approach would add a set of additional rows for *intermediate* materials (or forms of energy). There would be at least one such row for each classified commodity output of the extraction sectors (e.g., roundwood and ore concentrates) and perhaps also one row for processed material categories (e.g., primary iron, steel, primary aluminum, coke, gasoline, and propane). Each intermediate material would appear as a *negative* entry in the column for a sector where it is an output, and as a *positive* entry where it is an input. The negative and positive entries should balance (to zero) when converted into absolute quantities by multiplying the coefficients by total dollar outputs. This topic is discussed further in Chapter 6.

The waste output matrix W has one row for each pollutant that is separately accounted for in the system and one column for each industry sector (or commodity) of the main I–O table. The matrix element W_{ij} stands for residuals output, in physical units, per dollar value of the product of sector j. At least 60 residuals are being monitored in the United States, and the number could obviously be much larger if toxic chemicals or radioisotopes are added to the list, for instance.

The columns corresponding to extraction sectors (VII) indicate gross production of residuals (pretreatment) by extraction (or postextraction) processes. The coefficients would be in physical units (e.g., gram or kilogram) per dollar of output of the sector. Residuals explicitly accounted for would include mine tailings, acid mine drainage, silt from strip mining and plowing, agricultural runoff, stubble and harvest waste, pesticide residues, and stumps and branches. The columns corresponding to production sectors (VIII) indicate gross output of industrial wastes of all sorts. Consumption wastes by households can be associated with the "real estate" sector (INFORUM #168) or with "final private consumption."

The abatement sectors (segment IX) accept gross residuals outputs from other sectors as *inputs* (even if no processing is actually carried out). To prevent double counting, the entry for a given pollu-

tant—say BOD[†]—in the residuals output column corresponding to waste water treatment (one of the abatement sectors) will be a negative number. This is arrived at by taking the total of all gross BOD generated as a (negative) input to the treatment sector and adding the BOD that is discharged to the environment by the treatment sector itself. The difference is obviously negative and can be interpreted as the amount of BOD *eliminated* by the abatement sector.

Some wastes are eliminated by chemical transformation or by recycling them back into the economy as secondary materials. Recovered materials are just as much outputs of the pollution abatement and waste treatment sectors as "final" pollution, but they can conveniently be regarded as *negative inputs* to these same sectors and recorded as such in the resource input matrix (segment VI). They will also appear, of course, as *positive inputs* to the production sectors where they are used. This procedure, again, prevents double counting.

The resource input matrix R is a nonsquare matrix with n_1 rows and N columns (one for each sector of the I–O matrix). To obtain a summary vector of resource inputs to the economy, by type (in physical units), simply multiply R by the output vector X, namely,

$$r = RX \qquad (4.42)$$

where r is a vector of n_1 elements with one nonzero entry for each raw material extracted from the environment. Entries for recycled materials (scrap) and other intermediates will automatically be zero since such materials are produced and consumed entirely within the economy. While the units for each entry can be different, if all resource inputs that are measurable in terms of *mass* are given in the same units (e.g., tons), the sum of all such entries will approximate the total mass of materials that are extracted and used in connection with any economic activity.

The waste output matrix W, with n_2 rows and N columns is completely analogous, and a waste materials vector w is defined by the matrix multiplication:

$$w = WX \qquad (4.43)$$

As already explained, gross outputs of residuals are reduced by negative outputs from the abatement sectors (reflecting residuals elimination). Thus each

[†]Biological Oxygen Demand.

nonzero entry in w represents an outflow back to the environment. Again, if common physical units (e.g., tons) are used, the sum of all the vector elements in a column will approximate the total mass of materials returned to the environment (or sent to "treatment").

The two sums should, of course, be nearly equal, with a small difference between inputs and outputs due to net accumulation of capital goods (roads, structures, etc.). Unfortunately, the detailed balancing of materials inputs and outputs to the economy as described here has not yet been implemented in any large scale model. Most models have focussed on a subset of residuals outputs, either gross or net or both.

The I–O model implicitly assumes that the relationships between supply and demand for goods and services are in equilibrium. The specification of final demands, in particular, should also reflect a Pareto-optimal consumption pattern.

This is not intended to suggest that an equilibrium necessarily exists in the real world. On the contrary, it has been emphasized on several occasions that the environment provides important nonpriced goods and services, such as sunlight, rainfall, and waste assimilation capacity, that are subject to physical constraints but not to economic allocation mechanisms, due to indivisibilities and "public good" characteristics of the services in question.[56] Lacking such mechanisms, it is the business of politicians and legislators to design—hopefully with help from economists—devices for achieving as rational an allocation of resources as is possible by means of inherently clumsy tools such as standard setting and regulation. Without market pricing as a guide, of course, the political mechanism is awkward and very imperfect, but it is not totally ineffective. Of course, it is extremely difficult—perhaps even impossible, in principle—to simulate the political process in a model. The best we can do analytically, at present, is to incorporate demand for environmentally generated activities (such as pollution abatement) in the input–output model and treat the level of such demand as a policy variable of the model. While this procedure is hardly ideal, it is sufficiently realistic to be worthwhile.

In any event, to use the I–O model to investigate "alternative futures," the pattern of final demand cannot be considered as fixed, once and for all. On the contrary, it is clear that consumer demand for most goods and services depends on such variable

factors as income, unemployment level, price, and (for comsumer and producer durables) the existing and desired "stock." In general terms, the same factors would also determine the relative demand for environmental services and/or pollution abatement. Demand for producer goods, of course, depend on (lagged) demand for consumer goods, current capacity, depreciation rates, and the like.

Unfortunately, notwithstanding righteous dictums like "let the pollutor pay," the costs of removing pollutants must be borne by society and—ultimately—by final consumers. Although it is probably more efficient for pollution abatement costs to be borne directly by producers (and indirectly by consumers, through higher prices), the fact remains that if society chooses a high level of pollution abatement, sacrifices will have to be made somewhere.

The nature of the necessary tradeoffs can be seen by assuming an aggregate production function in which labor, capital, and natural resources are all intrinsically "scarce." Substitutions can be made, to some extent, between any pair of these factors of production, but none is freely or infinitely available. Hence to the extent that some capital, labor, and materials are diverted into a new pollution abatement sector,[†] less of these inputs will be available for producing conventional goods and services. Thus while pollution abatement *increases* consumer welfare to whatever degree it reduces environmental damages, it also *decreases* welfare to the extent that the economy can produce less of the other goods and services that people desire.[‡] The Pareto-optimal state would be one in which a single additional unit of societal expenditure on pollution abatement produces no net welfare increase at all: The marginal costs and benefits are exactly in balance.

That an optimal level of abatement exists (in principle) in most cases is assured by declining returns and rising costs. First, the marginal benefits—in terms of environmental damages avoided—*decrease* as one approaches closer and closer to a state of environmental "bliss" or zero discharge. Second, the marginal costs of eliminating each additional unit of pollution tends to *increase*, as one approaches the same limit. Thus clearly, the first few units of pollution eliminated may yield large benefits at little cost, but eventually the tables are turned and

---

[†]At least this is true in a society that employs its resources fully, or nearly so.

[‡]There are complications, of course; for instance, pollution abatement may also decrease costs for some producers.

incremental improvements beyond a certain point cost more than they are worth. That point is the optimum level of pollution treatment that we seek to identify.

To determine this point in practice, of course, requires two kinds of information that are difficult to obtain, and not contained in the I–O model or the resource/residual coefficients:

1. Damage functions (for each pollutant).
2. Cost functions (for each abatement sector).

The first category of information is scarce because the benefits in question are not priced on any marketplace. Economists are currently searching for satisfactory surrogates or imputation methods to deal with this problem.[77] The second category of information is troublesome for economists because it is not gathered systematically as a part of any of the statistical monitoring networks. Cost functions, in general, must be synthesized from engineering studies carried out on an industry-by-industry basis.[78]

If environmental damage functions were known, they could be utilized to construct appropriate final demand equations for pollution abatement services, in parallel to the final demand functions for other products and services. Cost functions, on the other hand, fix quantity of residuals eliminated per unit output of the abatement sector. Since output, in this case, is actually a measure of dollar revenues by the sector, which are costs to other sectors, the cost functions determine the magnitudes of the residual elimination coefficients.

Cost functions in the foregoing sense, of course, only define *direct* costs of abatement to the purchaser of the abatement service. However, the I–O model does permit us to compute the propagation of indirect costs to society as a whole, as is indicated shortly.

The basic scheme described heretofore in this section presumes a *given* cost of abatement activity and a given level of abatement. In principle, this might result from an optimization calculation, based on measured or imputed damage functions. In practice, it is more usual (and easier) to start from a politically mandated set of "standards" (or, more accurately, "goals") and assume that they will, in fact, be met.[†] Thus the problem of optimization is

[†]This is done for example in the Environmental Protection Agency's model SEAS. [67,68]

avoided in this model (but not in real life). SEAS, in its present form, can be used to explore alternative abatement scenarios and determine their macroeconomic impacts, in a crude way, through shifts in sectoral output, employment, and aggregate GNP.

However, except for Stone's simple tutorial example[64], only the recently developed UN World Model[62] has exploited the I–O model's ability to compute consumer price changes resulting from altered costs. In particular, it is possible to calculate the price effects of adding a pollution abatement sector to the economy, as is shown.

To illustrate how this procedure works, let us rewrite the basic I–O equation (4.3a) in matrix form.

$$\begin{bmatrix} I - A_{11} & \vdots & -A_{12} \\ ----- & - & ----- \\ -A_{21} & \vdots & I - A_{22} \end{bmatrix} \times \begin{bmatrix} X_1 \\ ---- \\ X_2 \end{bmatrix} = \begin{bmatrix} Y_1 \\ ---- \\ Y_2 \end{bmatrix} \quad (4.44)$$

where

A_{11} = the "old" I–O matrix (without abatement),
A_{12} = purchases by the abatement sector(s) (columns),
A_{21} = sales of the abatement sector(s) (rows),
A_{22} = sales of the abatement sector to itself,
X_1, Y_1 = the "old" output and final demand vectors without abatement,
X_2, Y_2 = total output and final demand for the abatement sectors per se.

$$\begin{bmatrix} V_1 \\ ---- \\ V_2 \end{bmatrix} \times \begin{bmatrix} I - A_{11} & \vdots & -A_{12} \\ ----- & - & ----- \\ -A_{21} & \vdots & I - A_{22} \end{bmatrix} = \begin{bmatrix} P_1 \\ ---- \\ P_2 \end{bmatrix} \quad (4.45)$$

Any shift in costs will tend to have an effect on prices. The direct cost of residuals control services to industry (or to consumers) is not the whole story. Since many industries are affected, the costs of purchased intermediate goods and services will also rise unevenly across the economy. The pattern of final consumption will also be affected. The easiest way to compute the overall impact of a pollution abatement program is to observe that, like other sectors, the residual treatment sector requires some labor and, therefore, has a characteristic value added by labor.

Consider two cases, without and with, residuals treatment activities. In the second case, we have a value-added vector with a nonzero element (V_2) and nonzero matrix elements A_{12}, A_{22}, and A_{21}, whereas in the first case, by definition, all of these terms vanish. Generally speaking, prices for all products will be different—and higher—in the second case (treatment) than in the first case (no treatment).

These equations can be used to determine the price consequences of changes in the cost structure of a given industry due to pollution abatement (or any other externally imposed constraint). Solving Equation (4.45) explicitly, it can be shown that *without* treatment

$$P_1 = A_{11}^{-1}V_1 \qquad (4.46)$$

whereas *with* waste treatment,

$$P_1^* = A_{11}^{-1}V_1 + A_{11}^{-1}A_{12}D\{A_{11}\}A_{21}A_{11}^{-1}V_1$$
$$+ A_{11}^{-1}A_{12}D\{A_{11}\}V_2 \qquad (4.47)$$

where $D\{A_{11}\}$ is the determinant of the matrix A_{11}.

Thus by substituting (4.46) in (4.47), we can write a general expression for the "new" price vector P_1^* in terms of the "old" one (for the old products):

$$P_1^* = \left[I + A_{11}^{-1}A_{12}D\{A_{11}\}A_{21}\right]P_1$$
$$+ A_{11}^{-1}A_{12}D\{A_{11}\}V_2 \qquad (4.48)$$

In addition, the price of the waste treatment services *per se* will be

$$P_2^* = A_{22}^{-1}V_2 - A_{22}^{-1}A_{12}P_1^* \qquad (4.49)$$

To assess the impact of these price changes on the consumer, the procedure would be to recompute the altered level of demand for each final good or service (other than pollution abatement). The reduction in consumption of nonenvironmental goods and services, because of abatement activity, is a measure of the social cost of the assumed level of residuals treatment. Lacking corresponding benefit measures, of course, it is not possible to determine analytically whether the criterion of optimality—marginal costs equal to marginal benefits—is met. Only the political process is capable of resolving such questions, rightly or wrongly.

REFERENCES

1. N. Spulber (ed.), "Foundation of Soviet Strategy for Economic Growth," *Selected Soviet Essays, 1924–1930*, University Press, Bloomington, IN, 1964.

2. A. Kaehler, "Die Theorie der Arbeiterfreisetzung durch die Maschine," *Eine gesamtwirtschaftliche Abhandlung des modernen Technisierungspezesses*, Hans Buske, Leipzig, 1933.

3. H. Haller, "Hans Peter's Contribution to Economics and Econometrics," *Econometrica*, **29** 84–85, (January 1961).

4. H. Loeb, *The Chart of Plenty, A Study of America's Product Capacity Based on the Findings of the NSPPC (National Survey of Product Capacity)*, Viking, New York, 1935.

5. H. Reichardt, *Kreislaufaspetke in der Oekonomik*, Mohr, Tuebingen, 1967.

6. W. W. Leontief, "Quantitative Input–Output Relations in the Economic System of the U. S.," *Rev. Econ. Stat.*, **18**, 105–125 (1936).

7. W. W. Leontief, "Interrelation of Prices, Output, Savings and Investment: A Study of Empirical Application of Economic Theory of General Interdependence," *Rev. Econ. Stat.*, **19**, 109–132 (August 1937).

8. W. W. Leontief, *The Structure of the American Economy, 1919–1929*, Oxford University Press, New York, 1941.

9. W. W. Leontief, *The Structure of the American Economy, 1919–1939*, second ed., Oxford University Press, New York, 1951.

10. W. W. Leontief, *Input–Output Economics*, Oxford University Press, New York, 1966.

11. W. W. Leontief (ed.), *Studies in the Structure of the American Economy*, Oxford University Press, New York, 1953.

12. W. W. Leontief, "Environmental Repercussions and the Economic Structure: An Input–Output Approach," *Rev. Econ. Stat.*, **52**, 262–271 (1970).

13. V. Riley and R. L. Allen, *Interindustry Economic Studies*, Bibliographic Reference Series No. 4, Johns Hopkins Press, Baltimore, 1955.

14. C. E. Taskier, *Input–Output Bibliography*, 1955–1960, United Nations, New York, 1961.

15. Netherlands Economic Institute, *Input–Output Relations: Proceedings of a Conference on Inter-Industrial Relations*, held at Driebergen, Holland, H. E. Stenfert Kroese, Leiden, 1953.

16. National Bureau of Economic Research, *Studies in Income and Wealth, Vol. 18, Input–Output Analysis: An Appraisal*, Princeton University Press, Princeton, N. J. 1955, also Technical Supplement, 1954.

17. T. Barna (ed.), *The Structural Interdependence of the Economy*, Proceedings of 1956 International Conference on Input–Output Analysis, Varenna, 1954.

18. T. Barna (ed.), *Structural Interdependence and Economic Development*, Proceedings of the International Conference on Input–Output Techniques, Geneva, September 1961, St. Martin's Press, London, 1963.

19. A. P. Carter and A. Brody (eds.), *Contributions to Input–Output Analysis*, Vols. I and II, North-Holland, Amsterdam, 1970.

20. A. Brody and A. P. Carter (eds.), *Input–Output Techniques*, North-Holland, Amsterdam, 1972.

21. J. Cornfield, W. D. Evans, and M. Hoffenberg, "Full Employment Patterns, 1950," *Month. Labor Rev.* **64**, 163–190, 420–432 (February and March 1947).

22. E. Glaser, "Interindustry Economics Research Program of the U. S. Government," in *Input–Output Relations, Conference on*

Inter-industrial Relations (edited by Netherland Economic Institute), H. E. Stenfert Kroese, Leiden, 1953, pp. 230–234.

23. J. L. Holley, "A Dynamic Model: I. Principles of Model Structure; II. Actual Model Structures and Numerical Results," *Econometrica*, **20**, 616–642 (October 1952); and **21**, 298–324 (April 1953).

24. R. A. Stone, *A Programme for Growth*, Vol. 1–8, University of Cambridge, Dept. of Applied Economics, 1962–1968.

25. M. K. Wood, "PARM—An Economic Programming Model," *Management Sci.*, **11**, 619–680 (May 1965).

26. C. Almon, Jr., M. B. Buckler, L. M. Horowitz, and T. C. Reimbold, *1985: Interindustry Forecasts of the American Economy*, Heath, Lexington, MA, 1974.

27. U. S. Bureau of Labor Statistics, *The Structure of the U.S. Economy in 1980 and 1985*, Bulletin 1981, Washington, DC, 1975.

28. R. A. Stone and G. Stone, *Social Accounting and Economic Models*, Bowes and Bowes, London, 1959.

29. W. H. Miernyk, *The Elements of Input–Output Analysis*, Random House, New York, 1965.

30. H. B. Chenery and P. G. Clark, *Interindustry Economics*, Wiley, New York, 1959.

31. U.S. Department of Commerce, Interindustry Economics Division, "The Input–Output Structure of the U. S. Economy: 1967," *Surv. Cur. Business* (February 1974).

32. *Input–Output Structure of the U.S. Economy, 1967*. Vols. 1–3, U. S. Department of Commerce, Washington, DC, 1974.

33. R. E. Kutscher, "The United States Economy in 1985," *Month. Labor Rev.*, (December 1973).

34. "Studies in Methods," *System of National Accounts*, Series F., No. 2, Rev. 3, United Nations, New York, 1968.

35. United Nations, Economic and Social Council, Statistical Classifications, Draft International Standard Classification of All Goods and Services (ICGS) Report of the Secretary-General—E/CN.3/457 (Part I), June 17, 1974.

36 "The Input–Output Structure of the Canadian Economy—1967," Vol. 1, Statistics Canada, Ottawa, August 1969.

37. W. A. Reardon, "An Input–Output Analysis of Energy Use Changes from 1947 and 1958 to 1963," Report to the Office of Science and Technology, Battelle-Northwest, Richland, WA, June 1972. Also "Input/Output Analysis of U. S. Energy Consumption," 6th International Conference on Input–Output Techniques, April 22–26, 1974.

38. A. M. Strout, "Technological Change and U. S. Energy Consumption 1939–1954," Ph. D. Thesis, University of Chicago, Chicago, IL, 1967.

39. R. Herendeen, "An Energy Input–Output Matrix for the U.S. 1963 Users Guide," CAC Document No. 69, University of Illinois, Center for Advanced Computation, Urbana, IL, 1973.

40. R. Herendeen and C. Bullard, "Energy Cost of Goods and Services, 1963 and 1967," University of Illinois, Center for Advanced Computation, Urbana, IL, 1974.

41. R. U. Ayres, T. Betlach, and C. Decker, "The Potential For Energy Conservation with particular Application to Personal Transportation and Space Heating," IRT-404-R prepared for Energy, Mines and Resources, Canada, Washington, DC, October 1975.

42. C. Almon, Jr., *The American Economy to 1975: An Interindustry Forecast*, Harper & Row, New York, 1966.

43. *Energy Model Data Base User Manual*, BNL 19200, Brookhaven National Laboratory, Upton, NY, 1975.

44. D. J. Behling, W. Marcuse, N. Swift, and R. G. Tessmer, Jr., "A Two-level Iterative Model for Estimating Interfuel Substitution Effects," BNL 19863, Brookhaven National Laboratory, Upton, NY, March 1975.

45. E. A. Hudson and D. W. Jorgenson, "The DRI Long Run Forecasting Model of the U. S. Economy," DRI, Inc., Cambridge, MA, February 1976.

46. Narkus-Kramer et al., design papers for SEAS, IR & T, mimeo, 1976.

47. J. Fisher, H. Hansberg, and L. Fischman, *Resources in America's Future*, Johns Hopkins University Press, Baltimore, 1960.

48. R. Ridker et al. (eds.), *Population, Resources and the Environment*, Vol. III, U. S. Commission on Population and America's Future, Research Reports, Washington, DC, 1972.

49. U. S. Bureau of Mines, *Mineral Facts and Problems*, Washington, DC, 1970.

50. R. U. Ayres, M. Narkus-Kramer, and A. L. Watson, "The Analysis of Resource Recovery and Waste Reduction Using SEAS," IRT-433-R, prepared for EPA, Washington, DC, May 1976.

51. T. Bingham and B. S. Lee, "An Analysis of the Materials and Natural Resource Requirements and Residuals Generation of PCE Items," Research Triangle Institute, prepared for EPA, June 1976.

52. R. U. Ayres and S. Noble, "Economic Impact of Mass Production of Alternative Low Emissions Automotive Power Systems," Appendix D, IRT-287, FR, prepared for U. S. Department of Transportation, Washington, DC, March 1973.

53. J. Cummings-Saxton et al., "The Economic Impact of Water Pollution Abatement Costs Upon the Chemical Processing Industries," prepared for the National Commission on Water Quality, IRT-413-R, March 1976.

54. U. S. Tariff Commission, "Synthetic Organic Chemicals," United States Production and Sales, 1972, Washington, DC, 1974. Production data is also given by the U. S. Census of Manufacturers, Industry Series (1972), Department of Commerce Publication MC72(2)-28.

55. For example: *Guide to the Chemical Marketing Industry*, Chas. H. Kline, New York 1974. *Chemical Engineering*, McGraw-Hill, New York. *Chemical Week*, McGraw-Hill, New York. *Chemical and Engineering News*, American Chemical Society, Washington, D.C.

56. O. Herfindahl and A.V. Kneese, *The Quality of the Environment*, Johns Hopkins University Press, Baltimore, 1965.

57. R. U. Ayres and A. V. Kneese, "Production, Consumption and Externalities," *Amer. Econ. Rev.*, **LIX** (7) (June 1969).

58. J. H. Cumberland, "A Regional Inter-industry Model for Analysis of Development Objectives," *Reg. Sci. Assoc. Papers*, *16* (1966).

59. J. H. Cumberland and J. R. Hibbs, "Alternative Future Environments, Some Economic Aspects," presented to the Institute of Management Sciences, National Bureau of Standards, March 1970.

60. J. H. Cumberland, "Environmental Implications of Regional Development," presented to Canadian Economics Association and the Canadian Council on Rural Development, Winnipeg, November 1970.

61. J. H. Cumberland, "Application of Input–Output Technique to the Analysis of Environmental Problems," prepared for 5th International Conference on Input–Output Techniques, Geneva, January 1971.

62. W. W. Leontief, "Environmental Implications and the Economic Structure: An Input–Output Approach," in S. Tsuru (ed.), *A Challenge to Social Scientists*, Asahi, Tokyo, 1970, reprinted in *Rev. Econ. Stat.*, **LII** (August 1970).

63. W. W. Leontief and D. Ford, "Air Pollution and the Economic Structure: Empirical Results of Input–Output Computations," 5th International Conference on Input–Output Techniques, Geneva, January 1971.

64. R. A. Stone, "The Evaluation of Pollution: Balancing Gains and Losses," *Minerva*, **X** (July 1972).

65. L. Ayres, I. Gutmanis, and A. Shapanka, "Environmental Implications of Technological and Economic Change for the United States, 1967–2000: An Input–Output Analysis," IRT-229-R, prepared for Resources for the Future, Washington, DC, 1971.

66. R. U. Ayres and I. Gutmanis, "Technological Change, Pollution and Treatment Cost Coefficients in Input–Output Analysis," Chapter 12, in R. G. Ridker (ed.), *Population, Resources and the Environment*, Vol. III, Commission on Population Growth and the American Future, Research Reports, 1972.

67. Environmental Protection Agency SEAS Documentation, unpublished 1974–75.

68. Ibid.

69. P. A. Victor, *Pollution: Economy and Environment*, Allen and Unwin, London, 1972.

70. R. Thoss, "A Generalized Input–Output Model for Residuals Management," 6th International Conference on Input–Output Techniques, Vienna, April 1974.

71. W. W. Leontief, "Structure of the World Economy: Outline of a Simple Input–Output Formulation," Nobel Lecture, Stockholm, Sweden, December 1973, reprinted in *Amer. Econ. Rev.*, **64** (5) December 1974.

72. A. Carter, W. Leontief, P. Petri, and J. Stern, "Study on the Impact of Prospective Environmental Issues and Policies on the International Development Strategy," Technical Report to United Nations Department of Economic and Social Affairs, Draft, May 1975, Final Report, 1976.

73. W. Isard, K. B. Bassett, C. L. Choguill, J. G. Furtado, R. M. Izumita, J. Kissin, R. H. Seyforth, and R. Tatlock, *Ecologic-Economic Analysis for Regional Development*, Harvard University Graduate School of Design, for the U. S. Department of Commerce, Environmental Science Services Administration, December, 1968, summarized in *Reg. Sci. Assoc. Papers*, **21** (1968).

74. W. E. Cooper and H. E. Koenig, "The Design and Management of Environmental Systems," Michigan State University, summarized in W. E. Cooper, T. C. Edens, H. E. Koenig, and A. Welden, "Toward Environmental Compatibility," Michigan State University, mimeo, 1973.

75. K. E. Watt, "Land Use, Energy Flow and Decision-Making in Human Society," University of California at Davis, mimeo, 1973.

76. W. O. Spofford, Jr., C. S. Russell, and R. A. Kelly, "Operational Problems in Large Scale Residual Management Models," in E. S. Mills (ed.), *Economic Analysis of Environmental Problems*, NBER, Columbia University Press, New York, 1975.

77. See K-G Maler and R. E. Wyzga, "Economic Measurement of Environmental Damage: A Technical Handbook," prepared for OECD Secretariat, Paris, mimeo, 1975.

78. See Environmental Protection Agency. "Cost of Clean Air," "Cost of Clean Water," and so on.

APPENDIX 4-A

DEFINITIONS OF SUBMATRICES OF THE SYSTEM OF NATIONAL ACCOUNTS (TABLE 4.1)

$T_{1.23}$ The holdings of financial assets by the institutional sectors at the beginning of the period of account.

$T_{1.24}$ The holdings of financial assets, issued by the country under study, by the rest of the world at the beginning of the period of account.

$T_{2.23}$ The holdings of net tangible assets by the institutional sectors at the beginning of the period of account. The resident economic agents from which the institutional sectors are built up hold between them all the tangible assets in the country in which they are resident; and, at the same time, the ownership of a tangible asset abroad is represented by the holding of a financial asset. As a consequence, the rest of the world is not represented in the system as holding tangible assets.

$T_{3.5}$ The inputs of commodities, reckoned at basic values, into the productive activity of industries.

$T_{3.6}$ The inputs of commodities, reckoned at basic values, into the productive activity of the producers of government services.

$T_{3.7}$ The imputs of commodities, reckoned at basic values, into the productive activity of producers of private nonprofit services to households. It is generally assumed that these inputs do not arise in the case of domestic services rendered on an individual basis.

$T_{3.8}$ Commodities, reckoned at basic values, entering into the consumption expenditure in the domestic market of all households, whether resident or not.

$T_{3.15}$ Additions to the stocks of commodities, reckoned at basic values, held by industries.

$T_{3.16}$ Additions to the stocks of commodities, reckoned at basic values, held by the producers of government services.

$T_{3.17}$ Commodities, reckoned at basic values, entering into the gross fixed capital formation of industries.

$T_{3.18}$ Commodities, reckoned at basic values, entering into the gross fixed capital formation of the producers of government services.

$T_{3.19}$ Commodities, reckoned at basic values, entering into the gross fixed capital formation of the producers of private nonprofit services to households.

$T_{3.24}$ Exports of commodities reckoned at basic values.

$T_{4.5}$ Commodity taxes, net, on the commodity inputs into the productive activity of industries. The sum $T_{3.5} + T_{4.5}$ represents these commodity inputs reckoned at producers' values.

$T_{4.6}$ Commodity taxes, net, on the commodity inputs into the productive activity of producers of government services.

$T_{4.7}$ Commodity taxes, net, on the commodity inputs of producers of private nonprofit services to households.

$T_{4.8}$ Commodity taxes, net, on commodities entering into household consumption expenditure in the domestic market.

$T_{4.15}$ Commodity taxes, net, on the commodities entering into the stocks of industries.

$T_{4.17}$ Commodity taxes, net, on the commodities entering into the gross fixed capital formation of industries.

$T_{4.18}$ Commodity taxes, net, on the commodities entering into the capital formation of producers of government services.

$T_{4.19}$ Commodity taxes, net, on the commodities entering into the capital formation of producers of private nonprofit services to households.

$T_{4.24}$ Commodity taxes, net, on exports of commodities.

$T_{5.3}$ Commodity outputs, reckoned at basic values, of industries.

$T_{5.4}$ Commodity taxes, net, on the outputs of industries. The sum $T_{5.3} + T_{5.4}$ represents the commodity outputs of industries reckoned at producers' values.

$T_{6.3}$ Commodity outputs, reckoned at basic values of the producers of government services.

$T_{6.8}$ Government services entering into household consumption expenditure in the domestic market.

$T_{6.9}$ Services produced for own use by government services.

$I_{7.3}$ Commodity outputs, reckoned at basic values, of producers of private nonprofit services to households.

$T_{7.8}$ Domestic services and private nonprofit services entering into household consumption expenditure in the domestic market.

$T_{7.10}$ Services produced for own use by private nonprofit services.

$T_{8.14}$ Final consumption expenditure on goods and services in the domestic market by resident households.

$T_{8.24}$ Final consumption expenditure on goods and services in the domestic market by nonresident households.

$T_{9.14}$ Final consumption expenditure by general government.

$T_{10.14}$ Final consumption expenditure by private nonprofit institutions.

$T_{11.3}$ Protective import duties.

$T_{11.4}$ Other import duties. In the standard accounts and tables, this entry includes all import duties and $T_{11.3}$ is suppressed.

$T_{11.5}$ Values added, that is compensation of employees, operating surpluses, provisions for the consumption of fixed capital and indirect taxes, net, in the productive activity of industries.

$T_{11.6}$ Values added in the productive activity of the producers of government services.

$T_{11.7}$ Values added in the productive activity of domestic services and the producers of private nonprofit services to households.

$T_{11.23}$ The negative of charges for the consumption of fixed capital.

$T_{12.11}$ Compensation of employees and operating surpluses classified by institutional sectors of origin.

$T_{13.12}$ Compensation of employees and operating surpluses arising in institutional sectors classified by component forms of income. For example, compensation of employees is divided between wages and salaries on the one hand and employers' contributions to social security and private pension funds, etc., on the other.

$T_{13.14}$ Current income transfers, including transfers of property income, paid out by institutional sectors (as sectors of receipt).

$T_{13.24}$ Current income transfers, including transfers of property income, paid out by the rest of the world.

$T_{14.11}$ Indirect taxes, net, paid to general government.

$T_{14.13}$ Gross receipts of income by the institutional sectors (as sectors of receipt).

$T_{15.20}$ Increases in stocks of industries.

$T_{16.23}$ The finance provided by the capital finance account of general government, of the increase in stocks of producers of government services.

$T_{17.20}$ Total gross fixed capital formation of industries.

$T_{18.23}$ The finance, provided by the capital finance account of general government, of gross fixed capital formation undertaken by producers of government services.

$T_{19.23}$ The finance, provided by the capital finance account of private nonprofit institutions, of gross fixed capital formation undertaken by the producers of private non-profit services to households.

$T_{20.23}$ The finance, provided by the capital finance accounts of the institutional sectors, of gross industrial capital formation (in stocks, and fixed assets) and the net purchases by these sectors of land and intangible assets other than financial assets.

$T_{22.23}$ Net acquisitions of financial assets by the institutional sectors.

$T_{22.24}$ Net acquisitions of financial assets, issued by the country under study, by the rest of the world.

$T_{23.1}$ The holdings of financial liabilities by the institutional sectors at the beginning of the period of account.

$T_{23.2}$ The net worths of the institutional sectors at the beginning of the period of account.

$T_{23.14}$ The saving of the institutional sectors.

$T_{23.21}$ Net receipts of capital transfers by the institutional sectors.

$T_{23.22}$ Net issues of financial liabilities by the institutional sectors.

$T_{23.25}$ Revaluations of financial liabilities held by the institutional sectors.

$T_{23.26}$ Revaluations of the net worths of the institutional sectors.

$T_{23.27}$ The holdings of financial liabilities by the institutional sectors at the end of the period of account.

$T_{23.28}$ The net worths of the institutional sectors at the end of the period of account.

$T_{24.1}$ Financial liabilities issued by the rest of the world and held by the institutional sectors at the beginning of the period of account.

$T_{24.2}$ The net worth of the rest of the world at the beginning of the period of account arising from its relationships with the country under study; that is

to say, the negative of the rest of the world's net indebtedness to that country.

$T_{24.3}$ Imports of commodities reckoned at c.i.f. values.

$T_{24.6}$ Direct expenditure abroad on goods and services by the producers of government services.

$T_{24.8}$ Final consumption expenditure abroad by resident households.

$T_{24.13}$ Gross receipts of income (whether distributed factor income or other current transfers) by the rest of the world from the country under study.

$T_{24.21}$ Net receipts of capital transfers by the rest of the world.

$T_{24.22}$ Net issues of financial liabilities, taken up by the country under study, by the rest of the world.

$T_{24.24}$ The rest of the world's balance of payments on current account with the country under study.

$T_{24.25}$ Revaluations of financial liabilities issued by the rest of the world and held by the country under study.

$T_{24.26}$ Revaluation of the net worth of the rest of the world arising from its relationships with the country under study.

$T_{24.27}$ Financial liabilities issued by the rest of the world and held by the institutional sectors at the end of the period of account.

$T_{24.28}$ The net worth of the rest of the world at the end of the period arising from its relationships with the country under study.

$T_{25.23}$ Revaluations of financial assets held by the institutional sectors.

$T_{25.24}$ Revaluations of financial assets issued by the country under study and held by the rest of the world.

$T_{26.23}$ Revaluations of net tangible assets held by the institutional sectors.

$T_{27.23}$ The holdings of financial assets by the institutional sectors at the end of the period of account.

$T_{27.24}$ The holdings of financial assets, issued by the country under study, by the rest of the world at the end of the period of account.

$T_{28.23}$ The holdings of net tangible assets by the institutional sectors at the end of the period of account.

Source: Reference 6.

APPENDIX 4-B

SIC from	to	BLS Sector	INFORUM Sector	Invest	INFORUM sector name (first time only)
1	99	—	—	—	
101		1	1	1	dairy farm products C
102		1	2	1	poultry and eggs C
103		1	3	1	meat, animals, and misc. livestock products C
104	199	—	—	—	
201		2	4	1	cotton E
202		2	5	1	grains E
203		2	6	1	tobacco E
204	207	2	7	1	fruits, vegetables, and other crops E
203	699	—	—	—	
711	731	4	10	1	agricultural, forestry, and fishery services
732	739	—	—	—	
741		3	8	1	forestry and fishery products E
742	799	—	—	—	
811		3	8	1	
812	821	—	—	—	
822	823	3	8	1	
824	841	—	—	—	
842	843	3	8	1	
844	849	—	—	—	
851		4	10	1	
852	859	—	—	—	
861		3	8	1	
862	899	—	—	—	
911	919	3	8	1	
921	988	—	—	—	
989		4	10	1	
991	999	—	—	—	
1011		5	11	2	iron ore E
1012	1019	5			
1021		6	12	2	copper ore E
1022	1029	6			
1031		7	13	2	other nonferrous metal ores E
1032	1041	7			
1042	1044	7	13	2	
1045	1049	7			
1051		7	13	2	
1052	1059	7			
1061		5			
1062		5	11	2	
1063		5			
1064		5	11	2	
1065	1068	5			
1069		5	11	2	
1071	1079	7			
1081		7	13	2	
1082	1091	7			
1092	1094	7	13	2	

E = extraction sectors
C = conversion sectors

128

SIC from	to	BLS Sector	INFORUM Sector	Invest	INFORUM sector name (first time only)
1095	1098	7			
1099		7	13	2	
1111	1112	8	14	2	coal mining E
1113	1199	8			
1211	1213	8	14	2	
1214	1299	8			
1311		9	15	3	crude petroleum and natural gas E
1312	1319	—	—	—	
1321		9	15	3	
1322	1379	—	—	—	
1381	1382	9	15	3	
1383	1388	9			
1389		9	15	3	
1391	1399	—	—	—	
1411		10	16	2	stone and clay mining E
1412	1421	10			
1422	1423	10	16	2	
1424	1428	10			
1429		10	16	2	
1431	1441	10			
1442		10	16	2	
1443	1445	10			
1446		10	16	2	
1447	1451	10			
1452	1456	10	16	2	
1457	1458	10			
1459		10	16	2	
1461	1469	—	—	—	
1471		11			
1472	1477	11	17	2	chemical fertilizer mining E
1478		11			
1479		11	17	2	
1481		10	16		
1482	1491	10			
1492	1499	10	16		
1501	1511	—	—	—	
1512		12	18	4	new construction
1513		13	13	4	
1514		14	18	4	
1515		15	18	4	
1516		16	18	4	
1517	1598	—	—	—	
1599		17	19		maintenance and repair construction
1601	1899	—	—	—	
1911		19	22	5	other ordnance
1912	1924	19			
1925		18	20	5	complete guided missiles
1926	1928	19			
1929		19	21	5	ammunition
1931		19	22	5	
1932	1939	19			
1941		19	22	5	
1942	1949	19			

SIC from	to	BLS Sector	INFORUM Sector	Invest	INFORUM sector name (first time only)
1951		19	22	5	
1952	1959	19			
1961		19	21	5	
1962	1989	19			
1991	1998	—	—	—	
1999			22	5	
2011		20	23	6	meat products C
2012		20			
2013		20	23	6	
2014		20			
2015		20	23	6	
2016	2019	20			
2021	2024	20	24	7	dairy products C
2025		20			
2026		20	24	7	
2027	2029	20			
2031	2037	20	25	8	canned and frozen foods
2038	2039	20			
2041	2046	20	26	9	grain mill products C
2047	2049	20			
2051	2052	20	27	10	bakery products
2053	2059	20			
2061	2063	20	28	11	sugar C
2064	2069	20			
2071	2073	20	29	12	confectionery products
2074	2081	20			
2082	2085	20	30	13	alcoholic beverages C
2086	2087	20	31	13	soft drinks and flavorings
2088	2089	20			
2091	2094	20	32	14	fats and oils C
2095		20	33	14	misc. food products
2096		20	32	14	
2097	2099	20	33	14	
2111		21	34	15	tobacco products
2112	2119	21			
2121		21	34	15	
2122	2129	21			
2131		21	34	15	
2132	2139	21			
2141		21	34	15	
2142	2199	21			
2211		22	35	16	broad and narrow fabrics C
2212	2219	22			
2221		22	35	16	
2222	2229	22			
2231		22	35	16	
2232	2239	22			
2241		22	35	16	
2242	2249	22			
2251	2254	24	38	19	knitting mills
2255		24			
2256		24	38	19	
2257	2258	24			

SIC from	to	BLS Sector	INFORUM Sector	Invest	INFORUM sector name (first time only)
2259		24	38	19	
2261	2262	22	35	16	
2263	2268	22			
2269		22	35	16	
2271	2272	23	36	17	floor coverings
2273	2278	23			
2279		23	36	17	
2281	2284	22	35	16	
2285	2289	22			
2291	2299	23	37	18	misc. textiles
2311	2389	25	39	20	apparel
2391	2397	26	40	21	household textiles
2393		26			
2399		26	40	21	
2411		27	41	22	lumber and wood products C
2412	2419	27			
2421		27	41	22	
2422	2425	27			
2426		27	41	22	
2427	2428	27			
2429		27	41	22	
2431		28	43	23	millwork and wood products
2432		28	42	23	veneer and plywood C
2433		28	43	23	
2434	2439	28			
2441	2443	28	44	24	wooden containers
2444		28			
2445		28	44	24	
2446	2449	28			
2451	2489	—	—	—	
2491		28	43	23	
2492	2498	28			
2499		28	43	23	
2511	2512	29	45	25	household furniture
2513		29			
2514	2515	29	45	25	
2516	2518	29			
2519		29	45	25	
2521	2522	30	46	25	other furniture
2523	2529	30			
2531		30	46	25	
2532	2539	30			
2541	2542	30	46	25	
2543	2589	30			
2591		30	46	25	
2592	2593	30			
2599		30	46	25	
2611		31	47	27	pulp mills C
2612	2619	31			
2621		31	48	27	paper and paperboard mills C
2622	2629	31			
2631		31	48	27	
2632	2639	31			

| SIC | | BLS | INFORUM | | INFORUM sector name |
from	to	Sector	Sector	Invest	(first time only)
2641	2643	31	49	27	paper products, NEC
2644		31	50	27	wall and building paper
2645	2647	31	49	27	
2648		31			
2649		31	49	27	
2651	2655	32	51	28	paperboard containers
2656	2659	32			
2661		31	50	27	
2662	2699	31			
2711		33	52	29	newspapers
2712	2719	33			
2721		33	53	30	books, periodicals, and misc. publishing
2722	2729	33			
2731	2732	33	54	30	other printing and services
2733	2739	33			
2741		33	58		misc. printing and publishing
2742	2749	33			
2751	2752	34	57		commercial printing
2753		34	58		
2754	2759	34			
2761		34	56		business forms, blank books
2762	2769	34			
2771		34	58		
2772	2781	34			
2782		34	56		
2783	2788	34			
2789	2791	34	58		
2792		34			
2793	2794	34	58		
2795	2799	34			
2811		35			
2812	2813	35	55	31	industrial chemicals C
2814		35			
2815	2816	35	55	31	
2817		35			
2818	2819	35	55	31	
2821		37	62	34	plastic materials and resins C
2822		37	63	34	synthetic rubber C
2823		38	64	34	cellulosic fibers
2824		38	65	34	noncellulosic fibers C
2825	2829	—	—	—	
2831		39	66	35	drugs
2832		39			
2833	2834	39	66	35	
2835	2839	39			
2841	2844	40	67	36	cleaning and toilet preparations
2845	2849	40			
2851		41	68	37	paints
2852	2859	41			
2861		35	61	33	misc. chemical products
2862	2869	35			
2871	2872	36	59	32	fertilizers
2873	2878	36			

| SIC | | BLS | INFORUM | | INFORUM sector name |
from	to	Sector	Sector	Invest	(first time only)
2379		36	60	32	pesticides and other agricultural chemicals
2881	2889	—	—	—	
2891	2893	35	61	33	
2894		35			
2895		35	61	33	
2896	2898	35			
2899		35	61	33	
2911		42	69	38	petroleum refining and related products C
2912	2914	42			
2915		42	70	38	heating oil C
2916	2949	42			
2931	2952	42	71	38	paving and asphalt C
2953	2991	42			
2992		42	69	38	
2993	2998	42			
2999		42	69	38	
3011		43	72	39	tires and inner tubes
3012	3019	43			
3021		43	73	40	rubber products
3022	3029	43			
3031		43	73	40	
3032	3068	43			
3069		43	73	40	
3071	3078	44			
3079		44	74	41	misc. plastic products
3081	3099	43			
3111		45	75	42	leather tanning and industrial leather products
3112	3119	45			
3121		45	75	42	
3122	3129	45			
3131		45	76	43	leather footwear
3132	3139	45			
3141	3142	45	76	43	
3143	3149	45			
3151		45	77	44	other leather products
3152	3159	45			
3161		45	77	44	
3162	3169	45			
3171	3172	45	77	44	
3173	3198	45			
3199		45	77	44	
3211		46	78	45	glass C
3212	3219	46			
3221		46	78	45	
3222	3228	46			
3229	3231	46	78	45	
3232	3239	46			
3241		47	81	45	cement, concrete, and gypsum C
3242	3249	47			
3251		47	79	45	structural clay products
3252		47			
3253		47	79	45	
3254		47			

| SIC | | BLS | INFORUM | | INFORUM sector name |
from	to	Sector	Sector	Invest	(first time only)
3255		47	79	45	
3256	3258	47			
3259		47	79	45	
3261	3264	48	80	45	pottery
3265	3268	48			
3269		48	80	45	
3271	3275	47	81	45	
3276	3279	47			
3281		48	82	45	other stone and clay products
3282	3289	48			
3291	3293	48	82	45	
3294		48			
3295	3297	48	82	45	
3293		48			
3299		48	82	45	
3311		49			
3212	3313	49	83	46	steel C
3314		49			
3315	3317	49	83	46	
3318	3319	49			
3321	3323	50	83	46	
3324	3329	50			
3331		51	84	47	copper C
3332		53	85	47	lead C
3333			86	47	zinc C
3334		52	87	47	aluminum C
3335	3338	—	—	—	
3339		53	88	47	primary nonferrous metals, nec C
3341		53	84	47	
3342	3349	53			
3351		54	84	47	
3352		55	87	47	
3353	3355	—	—	—	
3356		56	89	47	nonferrous rolling and drawing, nec
3357		56	90	47	nonferrous wire drawing and insulating
3358	3359	—	—	—	
3361		57	87	47	
3362		57	84	47	
3363	3368	57			
3369		57	91	47	nonferrous castings and forgings
3371	3389	—	—	—	
3391		50	83	46	
3392		57	91	47	
3393	3398	—	—	—	
3399		50	83	46	
3411		58	92	48	metal cans
3412	3419	58			
3421		62	98	52	cutlery, hand tools, and hardware
3422		62			
3423		62	98	52	
3424		62			
3425		62	98	52	
3426	3428	62			

| SIC | | BLS | INFORUM | | INFORUM sector name |
from	to	Sector	Sector	Invest	(first time only)
3429		62	98	52	
3431	3433	59	94	49	plumbing and heating equipment
3434	3439	59			
3441	3444	60	95	50	structural metal products
3445		60			
3446		60	95	50	
3447	3448	60			
3449		60	95	50	
3451	3452	61	96	51	screw machine products
3453	3459	61			
3461		61	97	51	metal stampings
3462	3469	61			
3471		62	101	52	other fabricated metal products, nec
3472	3478	62			
3479		62	101	52	
3481		62	99	52	misc. fabricated wire products
3482	3489	62			
3491		58	93	48	metal barrels, drums, and pails
3492	3493	62	101	52	
3494		62	100	52	valves, pipe fittings, and fabricated pipes
3495		62			
3496	3497	62	101	52	
3498		62	100	52	
3499		62	101	52	
3511		63	102	53	engines and turbines
3512	3518	63			
3519		63	102	53	
3521		64			
3522		64	103	54	farm machinery
3523	3529	64			
3531	3533	65	104	55	construction, mining, and oil field machines
3534	3537	66	105	55	materials handling machinery
3538	3539	—	—	—	
3541		67	106	56	machine tools, metal cutting
3542		67	107	56	machine tools, metal forming
3543		67			
3544	3545	67	108	56	other metal working machinery
3546	3547	67			
3548		67	108	56	
3549		67			
3551	3555	68	109	57	special industrial machinery
3556	3558	68			
3559		68	109	57	
3561		69	110	58	pumps, compressors, blowers, and fans
3562		69	111	58	ball and roller bearings
3563		69			
3564		69	110	58	
3565		69	113	58	industrial patterns
3566		69	112	58	power transmission equipment
3567		69	113	58	
3563		69			
3569		69	113	58	
3571		72			

SIC from	to	BLS Sector	INFORUM Sector	Invest	INFORUM sector name (first time only)
3572		72	115	60	other office machinery
3573		71	114	60	computers and related machines
3574		72	114	60	
3575		72			
3576		72	115	60	
3577	3578	72			
3579		72	115	60	
3581	3582	73	116	61	service industry machinery
3583	3584	73			
3585	3586	73	116	61	
3587	3588	73			
3589		73	116	61	
3591	3598	70			
3599		70	117	59	machine shop products
3611		74	118	62	electrical measuring instruments
3612	3613	74	119	62	transformers and switchgear
3614	3619	74			
3621		75	120	63	motors and generators
3622		75	121	63	industrial controls
3623	3624	75	122	63	welding apparatus and graphite products
3625	3628	75			
3629		75	122	63	
3631	3636	76	123	64	household appliances
3637	3638	76			
3639		76	123	64	
3641	3644	77	124	65	electric lighting and wiring equipment
3645	3649	77			
3651		78	125	66	radio and TV receiving
3652		78	126	66	phonograph records
3653	3659	78			
3661		79	127	67	communication equipment
3662		80	127	67	
3663	3669	—	—	—	
3671	3674	81	128	68	electronic components
3675	3678	81			
3679		81	128	68	
3681	3689	—	—	—	
3691	3692	82	129	69	batteries
3693		82	131	69	X-ray equipment and electrical equipment
3694		82	130	69	engine electrical equipment
3695	3698	82			
3699		82	131	69	
3711	3712	83	133	70	motor vehicles and parts
3713		83	132	70	truck, bus, and trailer bodies
3714		83	133	70	
3715		83	132	70	
3716		83			
3717		83	133	70	
3718	3719	83			
3721		84	134	71	
3722		84	135	71	
3723		84	136	71	
3724	3728	84			

| SIC | | BLS | INFORUM | | INFORUM sector name |
from	to	Sector	Sector	Invest	(first time only)
3729		84	136	71	
3731	3732	85	137	72	ship and boat building and repair
3733	3739	85			
3741	3742	86	138	73	railroad equipment
3743	3749	86			
3751		86	139	74	cycles and parts, transportation equipment
3752	3759	86			
3761	3789	—	—	—	
3791		87	140	74	trailer coaches
3792	3798	87			
3799		87	139	74	
3811		88	141	75	engineering and scientific instruments
3812	3819	88			
3821	3822	88	142	76	mechanical measuring devices
3823	3829	88			
3831		90	143	78	optical and ophthalmic goods
3832	3839	90			
3841	3843	89	144	77	medical and surgical instruments
3844	3849	89			
3851		90	143	78	
3852	3859	90			
3861		91	145	78	photographic equipment
3862	3869	91			
3871	3872	88	146	78	watches, clocks, and parts
3873	3879	88			
3881	3899	—	—	—	
3911	3914	92	147	79	jewelry and silverware
3915	3929	92			
3931		92	148	79	toys, sporting goods, musical instruments
3932	3939	92			
3941	3943	92	148	79	
3944	3948	92			
3949		92	148	79	
3951	3953	92	149	79	office supplies
3954		92			
3955		92	149	79	
3956	3959	92			
3961		92	147	79	
3962	3964	92	150	79	misc. manufacturing, nec
3965	3979	92			
3981	3984	92	150	79	
3985	3986	92			
3987	3988	92	150	79	
3989		92			
3991		92	150	79	
3992		92			
3993	3994	92	150	79	
3995		92			
3996		92	150	79	
3997	3999	92			
4011	4099	93	151	80	railroads
4111	4199	94	152	82	buses
4211	4299	95	153	81	trucking

| SIC | | BLS | INFORUM | | INFORUM sector name |
from	to	Sector	Sector	Invest	(first time only)
4301	4399	—	—	—	
4411	4499	96	154	82	water transportation
4511	4599	97	155	83	airlines
4611	4699	98	156	82	pipelines
4711		98			
4712		98	157	82	freight forwarding
4713	4719	98			
4721		98	157	82	
4722	4729	98			
4731	4739	95	153	81	
4741	4749	93	151	80	
4751	4779	98			
4781	4789	98	157	82	
4791	4799	98			
4811		99	158	85	telephone and telegraph
4812	4819	99			
4821		99	158	85	
4822	4829	99			
4831		100			
4832	4833	100	159	85	radio and TV broadcasting
4834	4839	100			
4841	4898	99			
4899		99	158	85	
4911		101	160	87	electric utilities C
4912	4919	101			
4921	4929	102	161	88	natural gas
4931			160	87	
4932		102	161	88	
4933	4938	—	—	—	
4939		101	160	87	
4941		103	162	88	water and sewer services
4942	4949	103			
4951		103	162	88	
4952	4959	103			
4961		103	162	88	
4962	4969	103			
4971		103	162	88	
4972	4979	103			
4981	4999	—	—	—	
5011	5099	104	163	84	wholesale trade
5101	5199	—	—	—	
5211	5299	105	164	84	retail trade
5311	5399	105	164	84	
5411	5499	105	164	84	
5511	5599	105	164	84	
5611	5699	105	164	84	
5711	5799	105	164	84	
5811	5899	105	164	84	
5911	5999	105	164	84	
6011	6099	106	165	86	credit agencies and brokers
6111	6199	106	165	86	
6211	6299	106	165	86	
6311	6399	107	166	86	insurance and broker's agents

| SIC | | BLS | INFORUM | | INFORUM sector name |
from	to	Sector	Sector	Invest	(first time only)
6411		107	166	86	
6412	6499	107			
6511	6519	109	168	86	real estate
6521	6529	109			
6531		109	168	86	
6532	6539	109			
6541		109	168	86	
6542	6549	109			
6551	6559	109	168	86	
6561	6599	109			
6611		109	168	86	
6612	6699	109			
6711	6799	106	165	86	
6801	6999	—	—	—	
7011	7099	110	169	56	hotel and lodging places
7101	7199	—	—	—	
7211	7299	111	170	86	personal and repair services
7311	7319	113	172	86	advertising
7321		112	171	86	business services
7322	7329	112			
7331	7332	112	171	86	
7333	7338	112			
7339	7342	112	171	86	
7343	7348	112			
7349	7351	112	171	86	
7352	7359	112			
7361		112	171	86	
7362	7389	112			
7391	7393	112	171	86	
7394	7395	112	171		
7396		112			
7397		112	171		
7398		112			
7399		112	171		
7401	7499	—	—	—	
7511	7599	115	173	86	auto repair
7611	7619	111			
7621	7631	111	170	86	
7632	7639	111			
7641		111	170	86	
7642	7691	111			
7602		111	170	86	
7693		111			
7694		111	171		
7695	7698	111			
7699		111	171		
7701	7799	—	—	—	
7811	7899	116	174	86	motion pictures and amusements
7911	7999	117	174	86	
8011		118	175	86	medical services
8012	8019	118			
8021		118	175	86	
8022	8029	118			

| SIC | | BLS | INFORUM | | INFORUM sector name |
from	to	Sector	Sector	Invest	(first time only)
8031		118	175	86	
8032	8039	118			
8041		118	175	86	
8042	8059	118			
8061		119	175	86	
8062	8069	119			
8071	8079	118	175	86	
8081	8089	118			
8091	8099	118	175	86	
8111	8199	114			
8211	8299	120	176	86	private schools and nonprofit organization
8301	8399	—	—	—	
8411	8499	121	176	86	
8501	8599	—	—	—	
8611	8699	121	176	86	
8701	8899	—	—	—	
8911	8919	114			
8921		121	176	86	
8922	8929	121			
8931	8999	114			
9001	9121	—	—	—	
9122		122	177		post office
9123		123	178		federal gov. enterprises
9124		124	178		
9125		125	180		state and local electric utilities
9126		126	181		directly allocated imports
9127		127	181		
9128		128	182		business travel
9129		129	183		office supplies
9131	9993	—	—	—	
9994			184		unimportant industry
9995			185		computer rental
9996	9996	—	—	—	
9999		103			

CHAPTER FIVE *Optimizing Materials/Energy Process Models*

with J. Cummings-Saxton, M. O. Stern, and R. W. Roig

It will be recalled that the "standard" assumption of interindustry input–output models is, in effect, that the inputs to each sector scale are in proportion to the outputs and remain constant (or change very slowly and deterministically) over time. Of course, this assumption is undesirable, since it does not permit substitution of resource inputs, technologies, or product mixes. The fact that it permits a "square" input–output matrix—one commodity per industry—is the usual, but not essential, reason for this restrictive assumption. (Even the nonsquare Canadian industry–commodity model assumes fixed input coefficients, as well as fixed market share or output coefficients.) However, in this chapter various means of generalizing the treatment of alternative technologies are explored.

The development in World War II of mathematical methods[†] of solving a set of static linear equations subject to linear (or, more recently, nonlinear) constraints on the variables[1,2] opened the possibility of relaxing the "unique and fixed technology" assumption and choosing among a variety of possibilities that one which maximizes over all output per unit of resource inputs. The multi-activity static optimizing model was formulated independently (in eco-

nomic terms) in 1951 by T. C. Koopmans[3] in the United States and L. V. Kantorovich[5] in the U.S.S.R. Koopmans never tried to implement the model for a realistic situation; he was satisfied to provide a number of theorems of purely theoretical significance. However, static optimization (e.g., linear programming) methods have been extensively applied to economic problems and process analysis since the early 1950s. Recent applications to technology choice, energy allocation, and residuals management are discussed later in this chapter.

The extension from static to dynamic programming techniques—enabling us to determine (at least in simple cases) an optimal path over time—has also led to the modern "theory of optimal controls." This development was also jointly fathered by an American, Richard Bellman,[5] and a group of Russians, L. S. Pontrjagin et al.[6] Dynamic optimizing models are discussed in some detail in Section 2.2.

The significance of the models described in this chapter is that they open the door to consideration of technology as an *explicit* factor of production. At the present time, alternate technologies must be described, and selections made from among them, in a static framework. (The theory of optimal controls has not, as yet, been applied to any problem of technological substitution even in a narrow context.

[†]These methods are familiarly called linear programming.

141

And the gap between micro-economics and macro-economics is, perhaps, wider in this area of analysis than in any other.) Optimizing models can only be made dynamic, at present, by strictly ad hoc methods, of which several are noted in passing. However, the emphasis, in the following, is on static optimization.

5.1 ACTIVITY ANALYSIS MODELS, GENERAL

The activity analysis model postulates an economy producing M (abstract) commodities,[†] but using some combination of $N > M$ different activities or processes[‡] to do so. Each activity is uniquely characterized by an output and a set of required inputs per unit output.

Each required input must be produced by some other activity. Thus, as in the input–output model, total production of any community is allocated between intermediate and final consumption, and the set of activities is interdependent as a result of these transactional relationships. For convenience, let us define terms once again, maintaining consistency with usage in the previous chapter.

The reader is referred to the partitioned system on Figure 4.8 and Equations (4.42) and (4.43). These linear equations determine resource inputs and residuals outputs for a total output vector X. In the ordinary (square) input–output model, the total output of commodities, X, is related to final demand as given by (4.3a):

$$Y = (1 - A)X \qquad (5.1)$$

where X and Y are presumed to be at the commodity level of disaggregation. In (5.1), each commodity is produced by only one sector, and each sector[§] produces only one commodity. These equations state that the final demand (for output from a given sector) is equal to the total output of that sector less the amount used as intermediate inputs to other sectors.

However, in the activity–analysis model, these restrictions are relaxed: Each commodity is produced

by more than one activity and each activity can produce more than one commodity. In place of a vector X of total outputs, we introduce a vector Z of activity levels. An element of this vector refers to the output (in dollars) of a particular commodity by a specified activity. In place of the matrix equation (5.1), a more general equation is introduced, namely,

$$Y = \left[Q^{(\mathrm{out})} - Q^{(\mathrm{in})} \right] Z \qquad (5.2)$$

where $Q^{(\mathrm{out})}$, $Q^{(\mathrm{in})}$ are rectangular matrices defining outputs and inputs (in \$) of each commodity per unit (\$) output of each activity, and Z is a vector giving output levels for each activity. Comparing (5.2) and (5.1) it can be seen that

$$\sum_k Q_{ik}^{(\mathrm{out})} Z_k = X_i \qquad (5.3)$$

or

$$Q^{(\mathrm{out})} Z = X \qquad (5.3a)$$

Similarly,

$$\sum_k Q_{ik}^{(\mathrm{in})} Z_k = \sum_j A_{ij} X_j (= X_i - Y_i) \qquad (5.4)$$

whence

$$Q^{(\mathrm{in})} Z = AX (= X - Y) \qquad (5.4a)$$

It is worthy of note that all the equations from (5.1) through (5.7) can also be expressed in physical terms. In keeping with the convention established in Chapter 4, upper case letters denote dollar units and lower case letters denote material units.

Equation (5.3) displays the allocation of production of the ith commodity among several activities or production processes. For instance, if the ith commodity refers to steel, it is well known that there are three major steel-making processes, namely,

Open hearth furnaces (OHF)
Basic oxygen furnaces (BOF)
Electric arc furnaces   (EAF)

The coefficient $Q_{\mathrm{OHF, ST}}^{(\mathrm{out})}$ defines the (\$) *output* of steel per (\$) output of OHF, and so forth. If we assume (for simplicity) that each of these processes produces *only* steel as a salable product—all other outputs being wastes—then the three coefficients

---

[†]The term "commodity" as used hereafter can also refer to a service.

[‡]Hereafter, the terms "activity" and "process" are used interchangeably. When more precise terminology is required, it is introduced.

[§]After "purification" of sectors (to eliminate by-products) are discussed in Section 4.1.

$Q_{i, \text{ST}}$ all have values of 1.0, and the total dollar output of steel X_{ST} is simply the sum of the dollar outputs by each of the three steel-producing activities

$$X_{\text{ST}} = Q^{(\text{out})}_{\text{OHF, ST}}Z_{\text{OHF}} + Q^{(\text{out})}_{\text{BOF, ST}}Z_{\text{BOF}} + Q^{(\text{out})}_{\text{EAF, ST}}Z_{\text{EAF}}$$

$$= Z_{\text{OHF}} + Z_{\text{BOF}} + Z_{\text{EAF}}$$

Obviously, if there were other salable by-products of steel furnaces (as there are in many industrial processes), then the $Q_{i, \text{ST}}$ coefficients would be less than unity.

Equation (5.4) similarly displays the allocation of industrial commodities as intermediate inputs to other activities. It is recalled that $A_{ij}X_j$ represents the fraction of the production of the jth sector (commodity) that is attributable to purchased inputs directly from the ith sector (commodity). Once again, suppose j is the index corresponding to steel (as a commodity), and suppose i, in this case, refers to pig iron (another commodity).[†] As noted previously, there are three processes, or activities, that produce steel as a product. But these processes do *not* utilize pig iron in the same proportions (in fact, electric arc furnaces depend mainly on scrap materials and use relatively little primary metal).

In this case, we have

$$A_{\text{PIG, ST}}X_{\text{ST}} = Q^{(\text{in})}_{\text{PIG, OHF}}Z_{\text{OHF}} + Q^{(\text{in})}_{\text{PIG, BOF}}Z_{\text{BOF}}$$

$$+ Q^{(\text{in})}_{\text{PIG, EAF}}Z_{\text{EAF}}$$

where $Q_{\text{PIG, OHF}}$ defines the ($) *input* of pig iron per ($) output of OHF, and so forth.

The $Q^{(\text{out})}$, $Q^{(\text{in})}$ matrices can obviously be combined into a single rectangular matrix, namely,

$$Q = Q^{(\text{out})} - Q^{(\text{in})} \qquad (5.5)$$

such that positive elements correspond to outputs and negative elements correspond to inputs.[‡] Matrix equation (5.2) now becomes simply:

$$Y = QZ \qquad (5.6a)$$

[†]Most input–output tables do not actually disaggregate the iron and steel industry to this degree, but that is irrelevant to the illustration.

[‡]Koopmans' original paper[3] asserts "the impossibility of the land of Cockaigne," that is, that no activity vector can have only positive and zero elements (or only negative and zero elements). In ordinary language, this postulate rules out activities with inputs but no outputs or outputs but no inputs. This is actually a weak (inequality) statement of the first law of thermodynamics.

This can also be expressed as a system of scalar equations

$$Y_i = \sum_k Q_{ik}Z_k \qquad (5.6)$$

The possible solutions of this set of equations are limited by a set of capacity constraints on the level of each activity:

$$Z_k \leqslant \bar{Z}_k \qquad (5.7)$$

An immediate extension of the basic activity analysis scheme is to augment the Q-matrix by defining energy and materials inputs and outputs in physical units. The scheme described in Section 4.5 is almost directly applicable, except that the square input–output matrix is replaced by a rectangular Q-matrix having one column for each activity. For each column-vector of the Q-matrix, we can associate a set of material/energy coefficients defining materials or energy inputs and outputs per unit ($) output of each activity. Total resource requirements and net residuals generated by any given configuration of activities can be calculated easily. The objective function to be maximized may also be a function of resource inputs and/or waste inputs, such that increasing social welfare is achieved (up to a point) by decreasing either or both of these quantities.

Another important constraint on possible solutions of Equations (5.6) is imposed by the laws of conservation of mass (and energy), discussed at length in Chapter 3. Here we must use material units. It must be recalled that $q\ (= q^{(\text{out})} - q^{(\text{in})})$ is a matrix of column vectors. Each vector defines all the physical inputs and outputs, including wastes, per unit of product. If these are measured in common units (of mass), then it is easy to see that the sum of all inputs and outputs must balance out to zero, namely,

$$\sum_i q_{ik} = 0 \qquad (5.8)$$

This equation holds for each activity or process k.

The materials balance principle can be applied on a more detailed level. Let γ_{il} be the lth elemental component (carbon, sulfur, iron, etc.) of the ith commodity, whence

$$x_i = \sum_l \gamma_{il} \qquad (5.9)$$

Since chemical elements are not (normally) transmuted into one another by industrial activities, the conservation rules hold individually for *each* element, namely,

$$\sum_i \gamma_{il} \, q_{ik} = 0 \qquad (5.10)$$

for all elements (indexed by *l*) and all activities (indexed by *k*). These equations (10) provide a powerful analytical tool for filling in gaps or reconciling discrepancies in data, as is discussed in the next chapter. Other constraints, for example, on residuals output or resource availability, can also be imposed. (Over a period of time, of course, such constraints can be modified by investment and depreciation, which converts the problem from a static to a dynamic one.)

For any well-behaved objective function $U(Y)$ depending on the pattern of final consumption, there exists at least one—and usually only one—set of activity levels corresponding to a vector Z that maximizes this function and is, by definition, an "efficient" or optimal mode of production. This optimal solution can be found, in principle, by enumeration and comparison or (more commonly) by use of a standard computerized linear programming algorithm. The details need not detain us here.

There have been a number of applications to economic problems on the micro-scale. A typical early use of linear programming would be to select the configuration of processes and products that maximize the profits of a petroleum refinery of fixed capacity for a given set of raw material and product prices.[7, 8] Problems of optimal industrial location and scheduling of additions to capacity also began to receive attention in the late 1950s and early 1960s.[9, 10] Some recent applications to energy, materials, and residuals management are discussed more explicitly in the next section.

Before proceeding to discuss actual models, however, I think it is worthwhile reiterating the distinction proposed earlier between the "old" paradigm of classical and neoclassical economics and the "new" paradigm in which the relations between goods and services and stocks and flows of real materials (and energy) are explicitly recognized and the consequences of physical laws that are applicable take precedence over strictly economic reasoning. Optimization models that treat resources, commodities, capital, labor, and residuals essentially in homogeneous monetary terms obviously belong to the former paradigm. The identification is more obscure where the physical nature of the production processes are not denied but merely disguised and/or aggregated to invisibility.

Linear programming, as applied to specific industries (such as petroleum refining), highlighted the very problems that prevented its application on a wider basis, namely, nonconstant returns-to-scale and strong interdependence between joint products using common production facilities. The petroleum refining industry is the standard example of an industry that can (and does) produce a wide range of different products, in a variety of "mixes," depending on seasonal demand and other factors. Joint (and variable) products are the rule, rather than the exception, in basic materials industries from metals to synthetics. This is because of the realities of physics, geology, chemistry, and biology. What nature has joined together by chemical bonds or biological structure, economic theory cannot safely assume apart.

Nor can the difficulty be resolved, in principle, by assuming a fixed relation between inputs and (joint) outputs based on the existence of a presumed "optimum" production process—unless all taxes, prices and wages are also simultaneously fixed. If these things were, indeed, fixed, there would be little point in undertaking economic analysis of any kind. But, in the real world, they are constantly in flux.

There are perhaps areas where the classical production picture (constant returns-to-scale, no joint products, etc.) may be adequate for some narrowly defined analytical purposes. In some cases, the adequacy of the linear approximation is presumed, but not particularly plausible—still less, demonstrated. However, the main purpose of this chapter is to discuss a model that attempts to overcome these deficiencies.

5.2 LINEAR PROGRAMMING MODELS FOR ENERGY ALLOCATION AND RESIDUALS MANAGEMENT

None of the models mentioned in this section are discussed in great detail. The interested reader is referred to the source documents in each case. The mathematics involved is comparatively well known, and the data base and outputs are far too voluminous to reproduce in detail. The major purpose of this short section is to provide a rough outline of the scope and capabilities of two important models and their places in the current pantheon.

BESOM[†]

One obvious candidate for large-scale application of activity analysis is to optimize the use of available fuels, energy transport, and conversion facilities in the economy. Obviously, many assumptions must be made in order to reduce this problem to tractable form, but a workable linear programming model (called BESOM) has been built by Brookhaven National Laboratory.[11, 12] The model consists of a specified energy flow *network*[‡] (shown graphically in Figure 5.1), and a set of constraint equations. The flow network displays pathways between seven sources of supply to 16 final demand types by way of a large number of alternative intermediate stages including fuel processing, transportation, electric power generation and production of "energy services" such as mechanical propulsion space heating and refrigeration. A list is shown in Table 5.1. Not all forms of energy are suitable for all uses, so many conceivable links are actually excluded. Nevertheless, there are 594 different specified pathways linking supply and demand nodes.

Clearly, the system can be reduced to a set of equations of the form (5.6) if we interpret y as a vector of demands for "energy services" to be exogenously specified, and z as a vector with 594 elements, one corresponding to each "chain" of possible links from supply to demand by way of alternate production processing and conversion steps. The coefficients of the q-matrix incorporate the structural relationships shown in Figure 5.1 together with efficiencies for each link. The realm of allowed solutions is limited by constraints of four different kinds, all of which can be expressed as linear inequalities, namely,

* *Supply.* Most energy sources are subject to short-term capacity limits, either physical or legal. (Total output of natural gas is limited by the number of producing wells and their pumping rates; consumption of natural gas by electric power generating plants is controlled by the FPC.)
* *Capacity.* Intermediate outputs are limited by

[†]BESOM, *Brookhaven Energy System Optimization Model.*
[‡]The data base for this network is available in Reference 13.

Table 5.1

Supply categories	Demand categories
Coal fired steam electric	Misc. electric base load
Coal fired steam electric, combined cycle	Misc. electric intermediate load
Oil fired steam electric	Misc. electric peak load
Oil fired steam electric, combined cycle	Electric storage
Oil fired gas turbine	Misc. thermal-low temperature
Gas fired steam electric	Misc. thermal-intermediate temperature
Total energy systems	Misc. thermal-high temperature
Light water reactor	Iron
Liquid metal fast breeder reactor	Petrochemicals
High temperature gas-cooled reactor	Space heat
Hydroelectric	Air conditioning
Geothermal	Water heat
Solar decentralized electric	Air transport
Pumped storage	Truck, bus
Oil, domestic	Rail
Oil, imported	Automobile
Oil, shale	
Natural gas	
Synthetic natural gas from oil	
Methane from coal	
Coal, underground	
Coal, stripmined	
Hydrogen from electrolysis	
Methanol from coal	
Hydrogen-coal	
Coal liquefaction	

throughput capacity of refineries, generating plants, and pipelines.
* *Peak load and cyclic requirements.* For demand categories that can utilize electricity or some other form of energy, a constraint is imposed to ensure that the overall relationship between off-peak base-load and peak-load for central power plants is held within a specified range. (The details are quite complex.)
* *Waste residuals.* Exogenous limits are imposed on total generation of certain types of waste residuals (such as SO_x, NO_x, and particulates). Waste loads are computed by means of "pollution" coefficients associated with each energy production or transformation activity.

The objective function for BESOM, in its present form, is (properly) not related to overall energy consumption, but rather with the simple sum of all costs associated with energy production, transformation, and use. If the unit costs are assumed to be constant, the objective function can be defined simply as the inner product

$$U = P_Z z \qquad (5.11)$$

where P_Z is the price (= cost) vector. This function can be minimized using linear programming methods.

The cost vector is constructed by simple summation over the chains, assuming a given unit cost for each step. The problem of scale is dealt with incompletely. In the case of electric power as already noted, output is explicitly allocated between baseload, intermediate-load, and peak-load demand, with the fractional division being parameters of the model. Both peak-load and base-load costs are constrained to cover capital recovery costs, whereas off-peak intermediate-load demand may be priced to recover only marginal fuel costs. It is noteworthy that both nonconstant returns to scale and joint product problems are dealt with explicitly in the case of electric power generation. Elsewhere in the model, however, these complexities are avoided and unit costs are apparently assumed to be independent of throughput.

BESOM is a static optimizing model, as emphasized already. It has been calibrated at the national level for 1969, and—after appropriate adjustment of constraint equations[†]—it reproduced the supply–demand mix and electric generating mix for that year with reasonable accuracy. The question arises whether a static optimum is meaningful in a time-dependent situation. Cherniavsky[12] suggests that a sort of quasidynamic optimum can be achieved by running it in a "sequential or time step mode where, at each step, additions to satisfy new demands and replacement of existing stock are optimized." In other words, if we assumed that the real world is actually close to the static optimum in a given year, it will *remain* close from year to year thereafter by optimizing short-term changes in capital stock to reflect short-term changes in demand.

Unfortunately, in the real world, capital investments require a long lead time to design, build, and

install, and—once in place—an even longer time normally elapses before a facility becomes totally obsolete, in the precise sense that its marginal cost of production (making no allowance for capital recovery) is higher than the market price for its product. For a central station generating plant, for instance, the preinstallation planning horizon ranges from 7 to 10 years, whereas the subsequent useful lifetime is seldom less than 30 years.

The breakeven price for operation or nonoperation of each plant varies over time, depending among other things, on the extent to which capital has been recovered through depreciation. Thus the assumption of fixed constant unit costs, independent of year or use, is erroneous in principle, as well as in practice. And, even if that assumption were a valid one, a true optimum for a general equilibrium system cannot be arrived at by considering only the current consumer interest (minimum price of energy), while ignoring the interest of producers and their employees, not to mention future generations.

The foregoing comment raises a question about the potential validity of successive static solutions (using a least-cost objective function) as an approximation to a dynamic optimization. However, it is not an argument against the use of programming methods *per se*. On the contrary, introduction of realistic economies-of-scale creates interdependences in the system that are otherwise ignored. This is obvious in the case of the central power generating station, where substantial economies-of-scale in the conversion of heat energy to electricity can (and do) outweigh distributional and other costs associated with centralization. The same principle applies in any situation where capital intensive facilities can be shared among several users, rather than being duplicated at higher cost per unit of capacity at different locations.

Sharing of facilities has its costs too, of course, including scheduling, warehousing, transportation, and the like. The marginal costs for the Lth user —or partner—are not the same as marginal costs for the first user, nor average costs for all users. But no matter which cost formula is applied, it is a function of the number of users, and their conditions of use.[†]

BESOM has been used to date mainly to evaluate alternative energy policy and R & D scenarios involving new technologies or energy source substitutions against a standard "base line" scenario that

---

[†]Since the model is totally defined by its network equations and constraints, the choice of constraints has a strong impact on the results obtained. However, there is nothing in the least illegitimate in tinkering with constraints to get a better "fit" to reality. This process is entirely analogous to tinkering with coefficients and equation forms in econometric models, though perhaps somewhat less systematic.

[†]We are concerned, in particular, about the amount of excess capacity *currently* available.

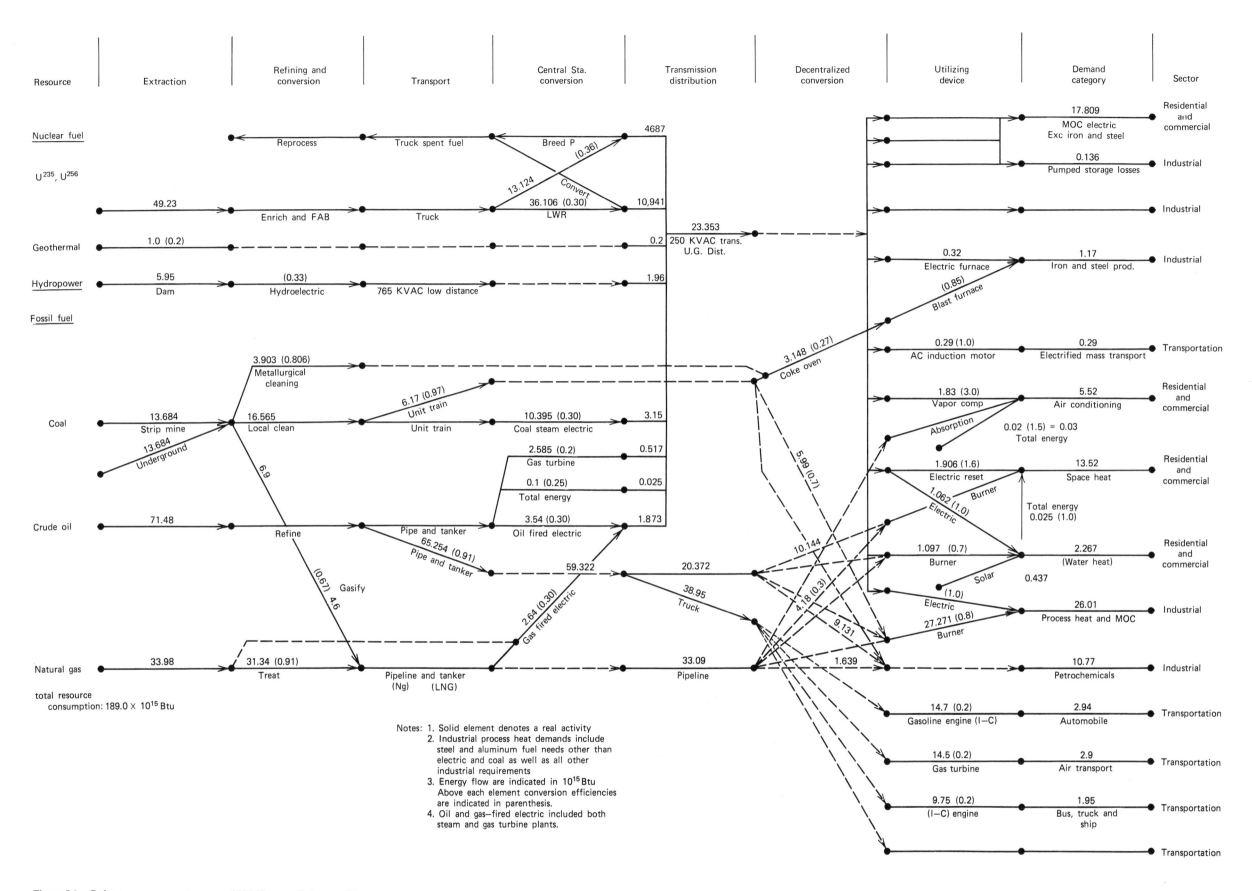

Figure 5.1. Reference energy system, year 2000 (Source: Reference 12).

has been established for the years 1985 and 2000. The optimization model has been linked to a 110-sector input–output table and a long-range macro-economic model by Behling et al.[14] The combined system operates iteratively. It begins with long-range aggregate trends (population, employment, GNP) and generates a set of final demands through time, which determine total industrial output and total energy requirements by sector. (The basic methodology is discussed in 4.2.) Demand for energy, in turn "drives" BESOM, which computes the optimal fuel mix and use of electric power. From the latter, a set of revised input vectors for the energy sectors are obtained for the next year. The changes from year-to-year generate new capital investment requirements, and contribute to final demand for the next year, and so forth.

Delaware Basin Model

A second important large-scale model for analyzing residuals management options in the Delaware Basin using linear programming modules (together with other models) has been developed over the past decade by Russell, Spofford, and Kelly at Resources for the Future, Inc.[15–17] The model was designed as, and still remains, primarily a tool for basic research. A secondary objective was to demonstrate the potential uses of such models either for on-line residuals management purpose or, more realistically, to provide off-line guidance to policy making or legislative bodies.[18]

Though it was conceived as a first step in a much grander design,[†] the Russell–Spofford model is already quite ambitious in scope. It combines a

[†]The model capabilities that one might ultimately require can be inferred from the following—admittedly visionary—paragraph from a Resources for the Future Monograph[19]:
The Administrator of the World Environment Control Authority sits at his desk. Along one wall of the huge room are real-time displays, processed by computer from satellite data, of developing atmospheric and ocean patterns, as well as the flow and quality conditions of the world's great river systems. In an instant, the Administrator can shift from real-time mode to simulation to test the larger effects of changes in emissions of material residuals and heat to water and atmosphere at control points generally corresponding to the locations of the world's great cities and the transport movements among them. In a few seconds the computer displays information in color code for various time periods—hourly, daily, or yearly phases at the Administrator's option. It automatically does this for current steady state and simulated future conditions of emissions, water flow regulation, and atmospheric conditions . . . .

linear programming model of industrial activities in the designated region with an environmental dispersion/transformation and ecosystem response model. An iterative feedback loop is also provided that compares output "state of the environment" (e.g., water or air quality in specified locations) with exogenously given standards. To the extent that given standards are not met, changes in the cost functions applicable to residuals producing industrial or municipal activities are imposed—"effluent charges" or penalties, in effect—and the system is rerun until an acceptable solution is found.

The overall computational scheme is illustrated in Figure 5.2.[16] The upper left-hand corner consists of a set of rectangular activity matrices representing major residuals-generating manufacturing activities such as petroleum refineries, steel mills, power plants, and paper plants, plus residuals-generating municipal activities such as sewage treatment plants and incinerators, as well as residual and commercial activities. The larger plants are considered as "point sources" located precisely by map coordinates in the eleven county Philadelphia region shown in Figure 5.3. The smaller industrial and commercial/household sources are treated as "area sources" assigned to grid squares.

Since the industrial plants and other residuals generators in the region already exist, the alternatives considered in the production/consumption part of the model pertain largely to operational modes and fuel choices affecting residuals discharge and disposal. A list of typical options by category of source is given in Table 5.2. Alternative materials processing or energy conversion technologies are not considered in the model.

The optimal choice of activities is formulated as a linear programming problem, using a "net benefit" objective function. In words, the objective is to maximize gross (regional) consumption (= output) net of all costs of treatment, transportation and disposal of residuals as well as some specified environmental engineering costs (such as aeration of river water to raise the level of dissolved oxygen) and any remaining damage costs. The objective function used here is conceptually more general than that defined for BESOM.

Some of the cost elements in the objective function depend on the residuals output in a strongly nonlinear fashion. This is true because, first, expenditures on pollution abatement tend to display sharply "declining returns" as discussed previously

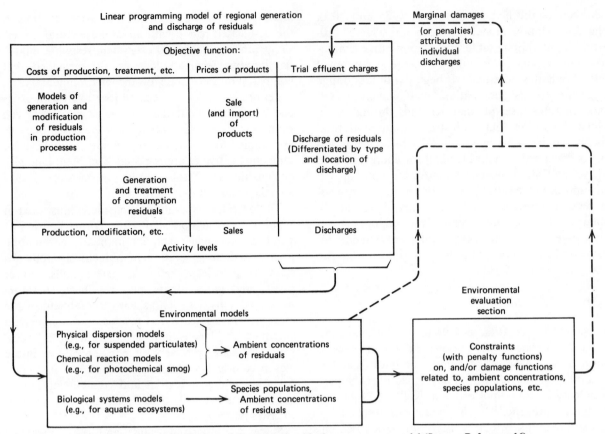

Figure 5.2. Schematic diagram of the regional residuals management model (Source: Reference 16).

(Section 4.5). That is to say, quite a lot of abatement may be purchased for the first increment of cost, but for every additional unit of pollution eliminated the marginal costs tend to rise even more rapidly. "Zero discharge," as a rough rule of thumb, would be infinitely costly.

But the problem of nonlinearities is more pervasive. Environmental dispersion and transformation processes, also, tend to be *essentially* nonlinear. This is notably true of the relationship between air pollutant discharge and the ambient level of photochemical smog. It is similarly true of the relationship between daily BOD[†] output and the dissolved oxygen level—or the number of fish likely to be found—in a river. In each case, the reason is fundamentally similar: The environmental reservoir can tolerate a certain amount of pollution with little harm done, but as the amount of waste increases, it rapidly overwhelms the natural assimilative mechanisms, perhaps even helping to destroy them.

[†]Biochemical Oxygen Demand is a rough measure of the organic content of waste materials.

Conceptually, these difficulties would seem to require a nonlinear constraint set as well as a nonlinear objective function. However, for computational convenience—since the model is quite large—only linear constraints were retained and certain nonlinear features were eliminated altogether. Other nonlinear constraints dealt with by converting them into incremental changes in costs[†] attributable to production or consumption activities. Finally, the nonlinear programming problem was converted into an iterative sequence of linear programming subproblems. Details are given in the basic documents, especially Reference 15.

The magnitude of the linear programming part of the model is indicated in Table 5.3. However, big as it is, this is not the biggest or most complex part of the model. The box in the lower left-hand corner of Figure 5.2 might be implemented in various ways,

[†]Technically, the technique of using "penalty" and "barrier" functions in place of nonlinear constraints is well known, see, for example, Zangwill.[20]

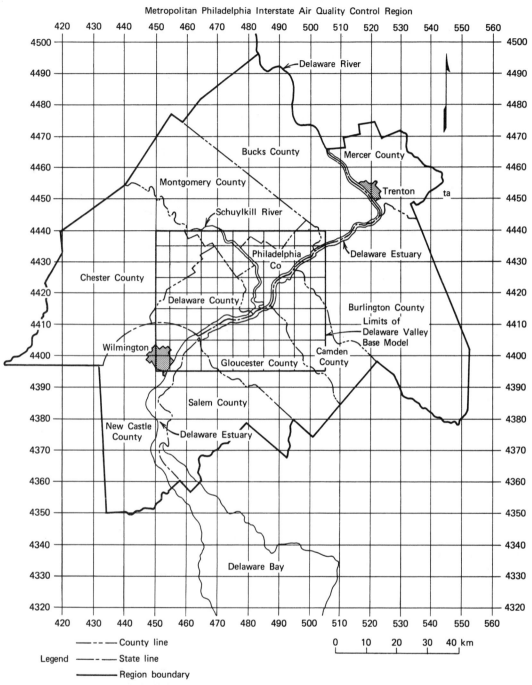

Figure 5.3. Delaware Valley Region (Source: Reference 16). Note: The grid is in kilometers and is based on the Universal Transverse Mercator Grid System (UTM).

Table 5.2   Residuals Management Options Available to the Various Types of Dischargers in the Region

Management option available	Primary residuals reduced	Secondary residual generated
Sugar refineries		
Partial or full reuse of flume water	BOD	Sludge
Secondary and tertiary wastewater treatment	BOD	Sludge
Cooling tower(s)	Heat	Heat is rejected to the atmosphere along with water vapor
Burn lower sulfur coal	SO_2	None
Electrostatic precipitators (3 alternative efficiencies; 90, 95, 98%	Particulates (fly ash)	Bottom ash
Sludge digestion and landfill	Sludge	The secondary residual here is digested sludge at a different location
Sludge dewatering and incineration	Sludge	Particulates
Dry cyclone; 90% efficiency	Particulates	Bottom ash
Petroleum refineries		
Secondary and tertiary treatment, and various reuse alternatives (cooling tower water makeup, desalinate water, boiler feedwater)	Nitrogen Phenols BOD	Sludge
Cooling tower(s)	Heat	Heat is rejected to the atmosphere along with water vapor
Burn lower sulfur fuel	SO_2	None
Refine lower sulfur crude	SO_2	None
Sell, rather than burn, certain high sulfur products (e.g., refinery coke)	SO_2	None
Cyclone collectors on catcracker catalyst regenerator (2 efficiencies; 70, 85%)	Particulates	Bottom ash
Electrostatic precipitator; 95% efficiency	Particulates	Bottom ash
Sell, rather than burn, high sulfur refinery coke	SO_2 Particulates	None
Sludge digestion and landfill	Sludge	The secondary residual here is digested sludge at a different location
Thermal power generating plants		
Cooling tower(s)	Heat	Heat is rejected to the atmosphere along with water vapor
Burn lower sulfur coal	SO_2	None

Table 5.2   (Continued)

Management option available	Primary residuals reduced	Secondary residual generated
Limestone injection-wet scrubber; 90%	SO_2	Slurry
Electrostatic precipitators; 90, 95, 98% efficiency	Particulates	Bottom ash
Settling pond; 90% efficiency	Slurry	Solid ash

Municipal sewage treatment plants

Secondary or tertiary wastewater treatment	BOD Nitrogen Phosphorus	Sludge
Sludge digestion, drying and landfill	Sludge	The secondary residual here is digested sludge at a different location
Sludge dewatering and incineration	Sludge	Particulates, bottom ash
Dry cyclone; 80, 95% efficiency	Particulates	Bottom ash

Municipal incinerators

Electrostatic precipitators (2 alternative efficiencies; 80, 95% efficiency)	Particulates	Bottom ash

Source: Reference 16.

but at present there are two physical dispersion/transformation models linked together and to the production/consumption/residuals generation (linear programming) module described previously. The first is a model to predict regional ambient concentrations of airborne pollutants, notably SO_x and particulates, developed by the U.S. Environmental Protection Agency. The second is a model to describe the flows of energy and nutrients between classes of organisms and predict several measures of water quality—notably dissolved oxygen (DO) level and fish population—in river water downstream from pollutant discharge points.[21] The river was subdivided into 21 distinct "reaches." Since this model is highly complex (and nonlinear), a somewhat simpler linearized water quality model of the Streeter–Phelps[22] type is now being incorporated in the system as an alternative.

In summary, the operation of the system is an iterative one, as follows. At the kth iteration the production/consumption activity (linear program-

ming) submodel is exercised and a set of residuals vectors (specified by location and environmental medium) is calculated. As inputs, this submodel uses a set of effluent charges (i.e., penalty functions) obtained as outputs from the $(k - 1)$st iteration. The newly calculated residuals become inputs to the ambient air quality and water quality models. A new set of ambient air and water quality levels is calculated and the results are passed on to the third submodel which evaluates the results, compares them with predetermined, "desired" standards and calculates a new set of penalty functions or effluent charges for the $(k + 1)$st iteration. When successive iterations result in no further significant improvements the iteration process is terminated.

A set of representative results, taken from Reference 17, is presented to give a flavor of the type of results the model can provide. Table 5.4 shows two alternative sets of air and water quality standards, "easy" (E) and "tight" (T), while Table 5.5 indicates the costs to 29 large industrial plants of meeting six

Table 5.3   Lower Delaware Valley Model: Residuals Generation and Discharge Modules

Module identification	Size of linear program			Description[a]	Percent extra cost constraints for the 57 political jurisdictions (except as noted)[c, e]
	Rows	Columns	Discharges		
MPSX 1	286	1649	130	petroleum refineries (7) steel mills (5) power plants (17)	57 electricity
MPSX 2	741	1482	114	home heat (57) commercial heat (57)	57 fuel 57 fuel[d]
MPSX 3	564	1854	157	"over 25 $\mu g/m^3$" dischargers (75)[b]	
MPSX 4	468	570	180	Delaware Estuary sewage treatment plants (36)	46 sewage disposal ($ per household per year)
MPSX 5	951	1914	88	paper plants (9) municipal incinerators (23) municipal solid residuals handling and disposal activities	57 solid residuals disposal
MPSX 6	229	395	117	Delaware Estuary industrial dischargers (22) in-stream aeration (22)	57 in-stream aeration (absolute extra cost per day)[f]
Total	3239	7864	786		

[a]The numbers in parentheses indicate the number of plants or activities that are included in the module. Some plants are included in more than one module; for example, twelve of the Delaware Estuary industrial wastewater dischargers in MPSX 6 are also represented by sulfur dioxide and/or particulate dischargers in MPSX 3.

[b]Industrial plants whose gaseous discharge result in maximum annual average ground level concentrations equal to or greater than 25 $\mu g/m^3$, except those in MPSX 1 (petroleum refineries, steel mills, and thermal power plants). The maximum annual average ground level concentrations (of sulfur dioxide and suspended particulates) were computed for each stack. For all stacks at the same x-y location (i.e., same plant), the maximum ground level concentrations were added together.

[c]The numbers in this column indicate the number of distributional constraints of a specified type that are incorporated in the model.

[d]The commercial heating requirements in this module are based on the differences between sulfur dioxide discharges from area sources of gaseous emissions and sulfur dioxide emissions from the home heating model.

[e]Only 46 of the 57 political jurisdictions discharge all or part of their sewage directly to the Delaware Estuary.

[f]The model currently reports the total regional absolute extra cost per day of in-stream aeration. This cost is then allocated equally among the 57 political jurisdictions. Any other distribution is also possible.

Source: Reference 17.

Table 5.4   Alternative Conditions Imposed on Regional Model for Initial Set of Production Runs

Feature involved	Alternative used		
Ambient air quality	Easy standards (E)	Stricter standards (T) (equivalent to federal primary standards)	
SO_2	$\leqslant 150\mu g/m^3$	$\leqslant 80\mu g/m^3$ [a]	
Suspended particulates	$\leqslant 120\mu g/m^3$	$\leqslant 75\mu g/m^3$	
Ambient water quality	Easy standards (E)	Stricter standards (T)	
Dissolved oxygen	$\geqslant 3.0$ mg/liter	$\geqslant 5.0$ mg/liter	
Algae	$\leqslant 3.0$ mg/liter	$\leqslant 2.0$ mg/liter	
Fish	$\geqslant 0.07$ mg/liter	$\geqslant 0.1$ mg/liter	
Solid waste disposal standards	Low quality landfill (L)	Medium quality landfill (M)	High quality landfill (H)
Required characteristics	open dump; no burning allowed	good quality sanitary landfill	good quality sanitary landfill with shredder, impervious layer to protect groundwater; wastewater treatment of leachate, aesthetic considerations such as fences, trees, and so on
Regional management alternatives	Instream aeration	Regional sewage treatment plant	
Alternatives available	—available in each reach	—two plant sites possible	

[a]Actually, for several locations the model could not meet the federal primary standards and for these the "strict" standards were relaxed to maintain feasibility.

Source: Reference 17.

Table 5.5   Costs to 29 Large Industrial Plants of Meeting Various Combinations of Air and Water Quality Standards[a] (costs in $1000 per day)

Water quality standards	Air quality standards	
	None	Easy[b]
None	0 [$\Delta = 19.6$]	$113 [$\Delta = 37$]
Easy[b]	$19.6 [$\Delta = 5.8$]	$150 [$\Delta = 17$]
Tight[b]	$24.4	$167

[a]The establishments are seven petroleum refineries, five steel mills, and 15 electric power plants.
[b]See Table 5.4 for definitions.

Source: Reference 17.

different combinations of these standards. For a discussion of the implications of these results, the reader is referred to the source documents.

Other Applications

I do not wish to leave the impression that the foregoing review of two applications of linear programming models to resource/environment problems is exhaustive. This is by no means the case. In particular, in addition to studies already cited, I feel obliged to mention a number of studies of optimal residuals management for particular industries carried out by Kneese, Bower, Löf Russell, and others over the last decade at Resources for the Future, Inc.[23–26] Another series of industry studies, focussing primarily on water pollution, has been undertaken by Russell Thompson and several colleagues at the University of Houston.[27-30] Thoss and his colleagues at the University of Muenster (Federal Republic of Germany) have built a generalized input–output model for residuals management on a regional basis, using a linear programming submodel to optimize the allocation of labor among cities in a region.[31, 32] Environmental quality and other constraints are imposed much as in the Russell–Spofford model. Another study of considerable interest has been carried out by Alan Manne at the International Institute of Applied Systems Analysis (IIASA).[33] Manne assumes a fixed demand pattern over time for energy in the United States and employs a sequence of linked linear programs to determine the optimal mix of electric-generating capacity, under the assumption that the date of commercial availability of breeder reactors is uncertain (i.e., random). Present decisions, not surprisingly, are found to be relatively insensitive to the timing of the breeder.[†]

5.3   MODELS FOR OPTIMAL CHOICE OF TECHNOLOGY

Each of the models considered until now has treated technology as an invariant or, at most, a peripheral element of the economic system. In BESOM,[‡] technological alternatives are considered explicitly at a level of detail that approaches realism, but BESOM is limited to energy related problems and focuses more on interfuel substitution than on technological choices *per se*. The Delaware Basin Model is concerned with residuals management options rather than technological choice. In the remainder of this chapter I discuss an optimizing model that was designed specifically for the study of technological options, taking into account the widest possible context including resource and environmental factors. The problem of forecasting technological change over time is discussed last.

The model in question is referred to as the materials–process–product model. Two versions are discussed, MPPM-I and MPPM-II. The first version was conceived by Ayres[38] and developed by Ayres,

Cummings-Saxton, and Stern.[39, 40] It retains some of the basic features of an activity analysis model, even though linear programming, as such, is not utilized. (BESOM could be regarded as a special case.) It incorporates explicit materials and energy balances, as well as realistic scale economics. Its chief limitation is that it assumes invariant ratios between material/energy inputs and invariant outputs. The second version, MPPM-II developed by Roig et al.[41] relaxes the latter constraint.

The Materials–Process–Product Model (I)

The first version of the materials–process–product model (MPPM-I) was developed primarily to facilitate quantitative analysis of material flows through the economy, where alternative but invariable technological possibilities exist. It is similar to, but somewhat more general than BESOM. Structurally, the model can be thought of as a collection of "nodes" connected by materials and energy flows. The materials/energy inputs and outputs of the economic processes composing an industry can be assembled in a set of columns with common definitions of the entries appearing in each numbered row, as illustrated by Table 5.6.[†] For example, suppose the industry consists of a final product (PVC bottles) and six basic intermediates—chlorine, ethylene, acetylene, ethylene dichloride, vinyl chloride mono-

Table 5.6   Typical Industry Process Matrix

	Processes			
	z_1	z_2	z_3	z_4
Inputs	$-q_{11}$	$-q_{12}$	$-q_{13}$	.
	$-q_{21}$	.	.	.
	$-q_{31}$	.	.	
Outputs	$+q_{41}$	.		
	$+q_{51}$			
	$+q_{61}$			
	$+q_{71}$			
	$+q_{81}$			

$\sum_{i=1}^{8} q_{ij} = 0$ by conservation of mass, where q_{ij} is the weight of material i participating in process j.

Source: Reference 34.

[†]The figures are presented as units of mass (lb) per \$ of output of the final product of the chain, namely PVC.

[†]There is also growing literature on the application of linear programming methods to optimal economic development strategy, which is outside the scope of this book.[34-37]

[‡]The Brookhaven Energy System Optimization Model is described in Section 5.2.

mer (VCM), and polyvinyl chloride (PVC)—produced by 32 different processes. The functional relationship (and flow) between the various materials and transformation processes is shown in Figure 5.4. The specific processes involved are listed in Table 5.7. The inputs consist of energy and primary raw materials, intermediate materials produced by preceding processes in the chain, and secondary materials, such as catalysts and drying agents. The outputs include the main product, all co-products or by-products, and the three waste streams—gaseous, liquid, and solid. If all mass flows are accounted for, the column sum for each process should be zero, as discussed previously, by the law of conservation of mass.

Before proceeding to a more detailed description of the model, it is worthwhile to explain the term "economic process," used throughout this section. In brief, an *economic process* is an industrial activity converting materials and energy from a low-value form to a higher-value form. It must meet two further criteria. First, the product(s) must be transportable and storable, and have an ascertainable market value (even if they are actually consumed within the same enterprise) so that separate accounts can be kept for the processing activity as distinct from other activities. Second, the activity must be carried out in the context of a cost-minimizing or a profit-maximizing enterprise. The first requirement rules out individual machines or unit operations that

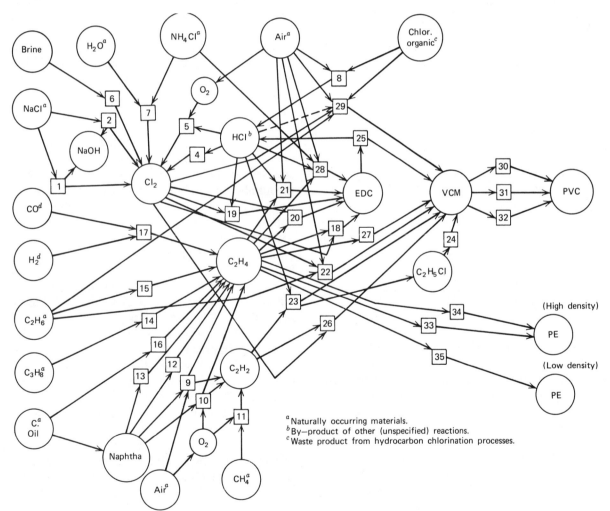

Figure 5.4. A network leading to polyvinyl chloride (PVC) and polyethylene (PE) (Source: Reference 35)

a—Naturally occurring materials.

b—By-products of other (unspecified) reactions.

c—Waste product from hydrocarbon chlorination processes.

Table 5.7   List of Processes

1. Cl_2 via salt electrolysis–mercury cell
2. Cl_2 via salt electrolysis—diaphragm cell
3. Cl_2 via HCl electrolysis
4. Cl_2 via HCl oxidation using $CuCl_2$ catalyst
5. Cl_2 via HCl oxidation using HNO_3 catalyst
6. Cl_2 via brine electrolysis and carbonation
7. Cl_2 via NH_4Cl oxidation—modified Solvay process
8. HCl from chlorinated HC waste
9. Acetylene/ethylene via Wulff process (naphthafeed)
10. Acetylene from naphtha by partial oxidation
11. Acetylene from methane by partial oxidation
12. Ethylene from light naphtha
13. Ethylene from full range naphtha
14. Ethylene from propane
15. Ethylene from ethane
16. Ethylene via autothermic cracking
17. Ethylene synthesis from CO/H_2
18. EDC via ethylene chlorination (liquid)
19. EDC via ethylene oxychlorination (liquid)
20. EDC via ethylene chlorination (gas)
21. EDC via ethylene oxychlorination (gas)
22. EDC via NH_4Cl oxychlorination
23. Ethylchloride from ethane/ethylene
24. VCM from ethylchloride
25. VCM from EDC pyrolysis
26. VCM from concentrated acetylene chlorination
27. VCM from acetylene/ethylene chlorination
28. VCM from ethane oxychlorination (transcat)
29. VCM from chlorinated HC waste
30. PVC from VCM—suspension polmerization
31. PVC from VCM—emulsion polymerization
32. PVC from VCM—bulk process
33. HDPE using metal oxide catalyst
34. HDPE using Ziegler catalyst
35. LDPE via tubular reactor

Source: Reference 35.

accept or produce materials in a transient or non-equilibrium state, or in a condition that requires a direct link to or from some other machine or operation. The second requirement rules out experimental, developmental or pilot plants or "backyard" or "standby" installations.

To visualize the working of the MPP model it is helpful to introduce a realistic example. Let us consider the intermediate chemical ethylene dichloride (EDC). At least five feasible processes are known, as shown in Table 5.7. Table 5.8 shows the material and energy inputs and outputs to these five processes, normalized to a unit (lb) of principal product. Needless to say, similar tables were compiled for each of the other processes involved in the complex of processes being analyzed.

The inputs and outputs are identified by number, taken from a master list of all materials (including wastes) that are consumed or produced by any of the 35 processes listed in Table 5.7. Only nonzero entries are explicitly shown, however.

As noted, the inputs and outputs shown in Table 5.8 refer to a physical unit of primary product (EDC). However, of the five feasible processes, only two are actually in use in the United States at present. These are, respectively, vapor-phase chlorination of ethylene (#20) and vapor-phase oxychlorination of ethylene (#21). Process #20 accounted for 65% of U.S. production in 1970 and process #21 accounted for the remainder. Taking into account the actual allocation of production by process, leads to Table 5.9, which is normalized to a unit output (lb) of the finished material, PVC. Table 5.10 indicates how actual production of selected intermediates in the manufacture of PVC were divided among various possible processes in 1970. Tables 5.8 to 5.10 are all adapted from Reference 39.

The following steps in the analysis are the critical ones. It is already obvious from the number of cross-linkages in the process-network Figure 5.4 that the number of available paths or chains leading to a finished material such as PVC is quite large. Nor can optimal choices for process technologies be made without considering all the other processes in such a chain—as well as all competing chains. Even if one leaves out of consideration (for the moment) the introduction of *new* process technologies, it is clear that a wide range of choices exists for the utilization of existing capacity. A "routing" program must first be devised, followed by a "linking" program. This, in turn, permits an efficient comparison and selection procedure. Together, these computer algorithms constitute the so-called process chain evaluation program or PCEP.

As noted already, an economic process is assumed to be characterized, in part, by its inputs and outputs of materials (and energy). Add labor and capital requirements and a complete production function is specified. What is true for a single process also holds for a *chain* of processes.[†] Thus the

[†]In the same way, an economic process is normally comprised of a network of subprocesses (or "unit operations").

Table 5.8 Unit Processes Normalized to 1 lb EDC Output

		1,2-Dichloroethane (EDC) production				
Code[a]	Material	Process 18[b] Product 31	Process 19[b] Product 31	Process 20 Product 31	Process 21 Product 31	Process 22[b] Product 31
Utilities						
1.	Power in kilowatt hours	0.02812	0.02659	0.01176	0.02553	0.02650
2.	Fuel in 1000 Btu	0	0	0	0	0.67446
3.	Steam in pounds	1.23939	0.54539	1.11963	0	0.21679
4.	Cooling water in gallons	6.19694	11.33603	11.35207	7.06324	12.21245
5.	Energy intensiveness in 1000 Btu	0	0	0	−0.31647	0
Primary raw materials						
7.	Water	0	0	0.30405	0	1.10712
8.	Oxygen from air	0.00336	0	0	0.18789	0.21727
9.	Oxygen, pure	0	0.18462	0	0	0
13.	Hydrogen chloride	0	0.78969	0	0.01729	0.36996
14.	Ammonium chloride	0	0	0	0	0.83372
Secondary raw materials, catalysts						
21.	Sodium hydroxide	0	0	0.00224	0	0
26.	Ferric chloride	0.00032	0.0005	0	0	0
Products						
29.	Chlorine	0.72089	0	0.72236	0	0
30.	Ethylene	0.28830	0.29658	0.30055	0.30373	0.32172
31.	1,2 Dichloroethane (EDC)	−1.00000	−1.00000	−1.00000	−1.00000	−1.00000
By-products						
36.	Ammonia	0	0	0	0	−0.19921
39.	Ammonium chloride	0	0	0	0	−0.17148
Air pollutants						
55.	Carbon monoxide	0	0	0	−0100125	−0.00154
56.	Nitrogen oxide	0	0	0	0	−0.00376
57.	Ammonia	0	0	0	0	−0.00188
59.	Hydrogen chloride	−0.00122	0	0	0	0
60.	Ethane	0	−0.00034	0	−0.00060	0
61.	Ethylene	−0.00400	−0.00284	−0.00007	−0.00279	−0.00272
62.	Ethylchloride	−0.00031	−0.00004	0	0	0
63.	Vinylchloride monomer (VCM)	0	−0.00001	−0.00122	−0.00871	−0.00951
64.	1,2-Dichloroethane (EDC)	−0.00047	−0.00178	−0.00169	0	−0.00193
65.	Dichloroethylenes	0	−0.00009	0	0	−0.00058
66.	1,1,2-Trichloroethane	0	0	0	0	−0.00012
Water pollutants						
67.	Water	0	−0.20283	−0.33384	−0.00224	0
69.	Sodium hydroxide	0	0	−0.00097	−0.00100	0
73.	Ferric chloride	−0.00032	0	0	0	0
74.	Sodium carbonate	0	0	0	0	−0.02770
80.	1,2-Dichloroethane (EDC)	0	−0.00223	−0.00482	−0.00197	0
82.	Tetrachloroethane	0	0	−0.00409	−0.00167	0

Table 5.8 (Continued)

| Code[a] | Material | 1,2-Dichloroethane (EDC) production | | | | |
		Process 18[b] Product 31	Process 19[b] Product 31	Process 20 Product 31	Process 21 Product 31	Process 22[b] Product 31
Contaminants						
86.	Water	0	− 0.00203	− 0.04644	− 0.19704	− 1.30676
88.	Ammonium nitrate	0	0	0	0	− 0.01149
89.	Sodium bicarbonate (Input)	0	0	0	0	0.04389
91.	Ethylchloride	− 0.00092	− 0.00204	0	0	0.00012
92.	1,2 Dichloroethane (EDC)	0	0	0	− 0.00148	− 0.01122
94.	Vinylchloride monomer	0	− 0.00028	0	0	− 0.01147
96.	Dichloroethylenes	− 0.00018	− 0.02011	0	− 0.00821	− 0.01178
99.	1,1,2-Trichloroethane	0	0	0	0	− 0.05795
100.	Chloral	0	− 0. − 1396	− 0.00467	0	0
101.	Perchloroethylene	0	− 0.00262	0	0	− 0.00200
102.	Chloral ammonia	0	0	0	0	− 0.00566
103.	Ethylene polyamines	0	0	0	0	− 0.00311
105.	Water (by-product)	0	0	0	0	− 0.03028

[a]Material codes refer to master list (not given here).
[b]Process not currently used in United States.

first step in evaluation by means of PCEP is to determine which chains of processes are feasible from a purely technical point of view. In other words, the first step in PCEP is to construct the basic network.

If we specify a certain desired product (say PVC) the conceptual problem is to identify all competing processes that will yield that particular product. Thence, working backward, the precursor materials required by those processes are determined from process chemistry and materials/energy balance considerations. For each such precursor the procedure is repeated: Competing production processes are identified, together with their input materials, and so on, until a complete materials–process network is specified, back to basic raw materials.

Of course, the network can also be arbitrarily truncated, as illustrated in Figure 5.4. Both the routing and linking procedures can obviously be carried out *ad hoc*, but in MPPM they have been programmed for a computer. This is an obvious advantage for purposes of updating: The routing and linking modules facilitate rapid and automated determination of the impact of the addition of a new process alternative to an existing set. A more fundamental purpose of the routing and linking modules is that they automatically classify all chains in the network into topologically similar sets. Such a grouping permits very efficient computation. I return to this point later.

It is additionally assumed for convenience that each precursor in each chain is dedicated entirely to that chain, and no other—all intermediate products produced in the chain being entirely consumed within it.[†] Then definite quantitative relationships can be determined between the final product of the chain and all purchased inputs to it, however many steps removed. Based on this same supposition, it is possible to determine the minimum scale of each precursor process along the chain. Thus to manufacture a single pound of PVC, at least 1.05 lb of vinyl chloride (monomer) and 1.235 lb of chlorine must have been produced somewhere. *Thus the scale of PVC demand dictates the scale of production of the intermediates.* (Of course, only a small proportion of all chlorine output goes into PVC, so chlorine production is actually on a much larger scale.)

These remarks indicate that—to some extent, at least—scale of output can be treated as an endogenous parameter of the model. If the MPPM covered the entire chemical industry—or, better yet, all the materials-processing sectors of the economy—the minimum scale of output for each material would be determined absolutely, given empirical data on capital and labor costs as a function of scale.[‡]

[†]More generally, it is consumed within the same network of chains.
[‡]The relationship between economies-of-scale and optimal choice of technology has been investigated recently for the engineering sectors by Westphal,[34, 35] Manne,[36] and Westphal and Rhee.[37]

Table 5.9 Materials/Energy Input–Output Vectors for Two Processes Actually in Use (normalized to 1 lb. PVC output)

		1,2-Dichloroethane (EDC) processes	
		Process 20 Product 31	Process 21 Product 31
Code[a]	Material		
Utilities			
1.	Power in kilowatt hours	0.01203	0.01406
3.	Steam in pounds	1.14533	0
4.	Cooling water in gallons	1.61261	3.89057
5.	Energy intensiveness in 1000 Btu	—	−0.17432
Primary raw materials			
7.	Water	0.31103	0
8.	Oxygen from Air	0	0.10349
13.	Hydrogen chloride	0	0.45018
Secondary raw materials, catalysts			
21.	Sodium hydroxide	0.00229	0
Products			
29.	Chlorine	0.73893	0
30.	Ethylene	0.30745	0.16730
31.	1,2-Dichloroethane (EDC)	−1.02295	−0.55082
By-products			
None			
Air pollutants			
55.	Carbon monoxide	0	−0.00069
60.	Ethane	0	−0.00033
61.	Ethylene	−0.00007	−0.00153
63.	Vinylchloride monomer (VCM)	−0.00124	−0.00480
64.	1,2-Dichloroethane (EDC)	−0.00173	0
Water pollutants			
67.	Water	−0.34151	−0.00123
69.	Sodium hydroxide	−0.00100	−0.00055
80.	1,2-Dichloroethane (EDC)	−0.00493	−0.00109
82.	Tetrachloroethane	−0.00418	−0.00092
Contaminants			
36.	Water	−0.04751	−0.10854
92.	1,2-Dichloroethane (EDC)	0	−0.00081
96.	Dichloroethylenes	0	−0.00452
100.	Chloral	0.00477	0

[a]Material codes refer to a master list (not given here).

Taking smaller and smaller subsets of the materials-processing system as a whole, naturally increases the uncertainties, as "imports" and "exports" become progressively more important. In the case of PVC, several of the key intermediates—notably chlorine and ethylene—dedicate very little of their output to PVC and export most of it to other products. Alternatively, we can simply regard these large volume precursor materials as imports from outside the system, whose scale of output, and market price will be essentially unaffected by the presence of absence of demand from PVC.

Having determined consolidated material/energy requirements for a "chain," and the minimum scale of production for each precursor—subject to the standard assumption of fixed inputs and outputs—it is quite straightforward to calculate the consolidated minimum capital and labor requirements for each

Table 5.10 Allocation of Production by Process, PVC, and Intermediates

Process # (See Table 5.8)	Material					
	Chlorine	Ethylene	1,2-Dichloro-ethane (EDC)	Ethyl chloride	Vinyl chloride monomer (VCM)	Polyvinyl chloride (PVC)
1	0.28					
2	0.66					
3	0.02					
4	—					
5	—					
6	0.01					
7	—					
Other [a]	0.03					
12		—				
13		—				
14		0.41				
15		0.44				
Other		0.15				
18			—			
19			—			
20			0.65			
21			0.35			
22			—			
23						
Other				0.95		
				0.05		
24					—	
25					0.92	
26					0.06	
27					0.02	
28					—	
29					—	
30						0.81
31						0.10
32						0.05
Other						0.04
1970 Production (lb × 10⁹) allocated to PVC	2.25	1.45	4.8	0	3.2	3.05

[a]Other refers to other processes for making that product.

link and, by aggregation, for the whole chain. The effective production function for the product in question is thus determined. (In engineering practice, utilities are generally considered separately, and the MPPM adheres to this convention.) Capital requirements can be, and are, broken down into fixed investment I_F, auxiliary investment I_A, and working capital, plus land I_W. Labor can be subdivided into direct production labor and indirect (overhead and sales) labor, fringe benefits, and the like.

To carry out the procedure described here, the process chain evaluation program (PCEP) must be supplied with a set of appropriate "dossiers" on each precursor process. Each process dossier contains a catalog of information on material and energy requirements per unit of output (i.e., input coefficients), plus corresponding coefficients to determine by-products and waste products. In addition, the process dossiers include data on capital and labor requirements, as well as scale factors and financial data.

In effect, the PCEP system requires input data on each of the various categories of capital investment (e.g., fixed investment, auxiliary investment, etc.)

and labor, as a function of plant capacity and capacity utilization levels. This information could be supplied internally by a fully detailed separate "model" of the industry. Or it could be gathered from empirical research and stored in tabular form or in the form of statistically fitted econometric equations. In MPPM-I, the latter alternative has been preferred, since a number of semiempirical relationships—known as "rules of thumb" to professional engineers—are already available in the standard literature.[†42-49]

For purposes of illustration, consider the method of Zevnik and Buchanan.[49] They developed the following empirical formulae for fixed capital investment, $I_{FC} = I_F + I_A$

for $q_D < 10,000,000$ lb/yr—

$$I_{FC} = 0.01 F_C q_D^{0.5} N \frac{I_{ENR}}{300} \qquad (5.12a)$$

for $q_D \geq 10,000,000$ lb/yr—

$$I_{FC} = 0.008 F_C q_D^{0.6} N \frac{I_{ENR}}{300} \qquad (5.12b)$$

In this case, the *Engineering News Record* Index has been used to express the time variation of construction costs, and the capacity exponents are assumed to be 0.5 and 0.6 for the two capacity ranges. The other parameters in the equations are:

$$F_C : \text{process complexity factor}$$

and

$$N : \text{number of operating units}$$

The process complexity factor is defined by the equation

$$F_C \equiv 2 \times 10^k \qquad (5.13a)$$

where

$$k \equiv F_t + F_p + F_a \qquad (5.13b)$$

F_t is a process temperature factor given by

$$F_t = 1.8 \times 10^{-4}(T - 27) \qquad T \geq 27°C \quad (5.14a)$$

$$= 2.0 \times 10^{-3}(27 - T) \qquad T < 27°C \quad (5.14b)$$

where T is the maximum process temperature (degrees Celsius). F_p is a process pressure factor

$$F_p = 0.1 \log p \qquad p \geq 1 \text{ atm} \qquad (5.15a)$$

$$= 0.1 \log \frac{1}{p} \qquad p < 1 \text{ atm} \qquad (5.15b)$$

where p is a maximum process pressure (atmospheres). F_a is a process alloy factor that depends upon type of material

$F_a = 0$ for cast iron, carbon steel, wood
   $= 0.1$ for aluminum, copper, brass, type 400 stainless steel
   $= 0.2$ for type 300 stainless steel, nickel, Monel, and Inconel alloys
   $= 0.3$ for Hastelloy alloys
   $= 0.4$ for precious metals

Although it is preferable to have greater flexibility in the capacity exponent, the expressions of Zevnik and Buchanan furnish a reasonable approximation for fixed capital investment.

Initial investment decisions are normally based on a comparison of the discounted present value of estimated future returns on alternative investment opportunities. However, once the investment is made and the plant is built, decisions tend to be made on the basis of current cash flow considerations. Thus in a static equilibrium situation capital investment must be treated as an (equivalent) annual financial cost of operation, in parallel with labor costs, material costs, utilities, and taxes. As a practical matter, the equivalence between front-end investment and annual financial costs depends on the form of the investment (equity or debt) and on the applicable corporate and other tax laws. The MPPM incorporates a fairly comprehensive tax submodel, which permits a user to assess the impact of alternative types of taxation on the optimal choice of technology. A description of the "annualization" procedure used in MPPM-I is included as Appendix 5-A.

While implicit in what has already been said, it is noteworthy that the PCEP computes detailed materials/energy balances for each chain. In particular, it computes aggregate *energy* consumption for each process sequence and quantifies all forms of waste energy. It also computes aggregate *residuals output* for each chain, by discharge medium (air, water, solid). Indeed, it was to perform these tasks in a process-specific manner that the model was initially proposed.[38]

---

†Obviously outdated cost information cannot be used without appropriate adjustment for price inflation. The ENR Construction Cost Index published in *Engineering News Record* and the CE Plant Cost Index published in *Chemical Engineering* (among others) can be used for this purpose.

The last step carried out in the process chain evaluation program is, of course, the selection of a "best" technology. This can only be done, of course, for a specified objective function. In PCEP, the objective function is completely arbitrary. It can be any well-behaved function, linear or nonlinear, of the extensive variables that describe a process—notably materials and energy inputs or outputs, labor, and capital—since a process chain is described by exactly the same *set* of variables. We can seek to minimize venture cost, working capital, direct labor, energy input, or pollution output—or any desired combination of these objectives.

This flexibility is possible because the PCEP algorithm proceeds to identify a global optimum by a systematic sequential comparison and partial-ordering procedure in which there are never more than a few competing alternatives at any stage.

In MPPM-I, this procedure hinges upon the fact that each product is produced by no more than three or four primary processes, each of which—in turn—can be evaluated as though its inputs were all purchased on the open market. Thus starting at the "end point" (PVC), the first stage comparison involves only three polymerization processes, all using the same inputs (VCM). For a given set of exogenous constraints, market prices and choice of utility function the optimum choice is unambiguous.

The overall optimization problem is now reduced to finding the best chain leading up to VCM. Again, we need only look one step back. There are six alternatives to examine, each one utilizing input materials (not the same ones, this time) that are available in the open market. Again, the comparison is unambiguous providing one can devise an appropriate convention for by-products.[†] (In this case, it happens that the viability of one of the processes depends upon the salability of the by-product HCl.) And so the program continues until arriving at raw materials or being commanded to stop and look no further.

To explore the sensitivity of "optimal" choices under alternate assumptions, MPPM-I has been used to rank-order 62 chains to produce PVC bottles under five different sets of constraints roughly simulating conditions in different countries. This exercise was carried out with a subset of the process

network shown in Figure 5.4, including only those processes that are currently in use in various countries. The five different cases are described in Table 5.11. Four of them correspond roughly to conditions in countries at various stages of development. The fifth was introduced to explore a possible demand reduction measure, though it also reflects quite well the energy price increase that actually occurred in late 1973. Obviously "zero labor cost" is not realistic, in practice—even as a limiting case—since there is always a shortage of skilled labor, but it allows us to identify the process chains that are most labor-intensive. The capital penalties are probably understated, if anything, for developing countries.

For the sensitivity exercise, the objective function (to be minimized) was taken to be "venture cost" of production of PVC. The entire set of 62 chains were ranked by this measure, for each of the five "scenarios." As might be expected, the rank-orders varied quite a bit from one scenario to another, although one chain showed up at the top of the list in three different cases (#1, 3, 4) and was no worse than sixth-ranked in any case (#2).

Another chain, ranked second in the "standard" case (#1) and dropped very low in the rankings when labor costs were reduced, but still ranked in second place for the energy tax scenario (#5). However, the chain that was ranked first in scenario #5 was ranked only 10th in case #1 and as low as 39th (out of 62) in case #2.

Table 5.11 Representative Variations in Energy, Labor, and Capital Costs

Case no.	Type country[a]	Parameter specification
1	U.S. industry	Base case, refer to Chapter 4
2	Limiting case less developed country (LDC)	Zero labor cost
3	More realistic LDC	Labor 10% U.S., capital Penalty 20%
4	Intermediate case developing country	Labor 25% U.S., capital Penalty 10%
5	U.S. industry, with internalized social costs	Energy use tax, $0.002/MBtu

[a]An example of a "limiting case," LDC might be Bangladesh; a more realistic LDC might be India or Indonesia; an intermediate case developing country might be Mexico.

Source: Reference 36

[†]The fact that some intermediates are also available as by-products introduces occasional difficulties for which no general universal solution has yet been discovered. Such cases have been treated largely *ad hoc* up to the present.

As one might expect, the optimal chains for the various scenarios result in quite different energy and materials requirements, waste heat output, pollution loads and other characteristics. The results are summarized in Figure 5.5. Note that only three of the five cases led to different optimum chains—the highest ranking chain for the "standard" case (#1) also ranked first for cases #3 and 4, as remarked already. From Figure 5.5, for instance, it can be seen that the optimum chain in case #2 (zero labor cost)

also consumes nearly 40% more energy than the standard case (#1). On the other hand, in case #5 where the energy tax was imposed, the optimum chain is one that reduces net energy consumption by about 12% as compared to the standard case.

The Materials–Process–Product Model (II)

It was pointed out in the foregoing discussion that the key limitation of MPPM-I is the assumption that

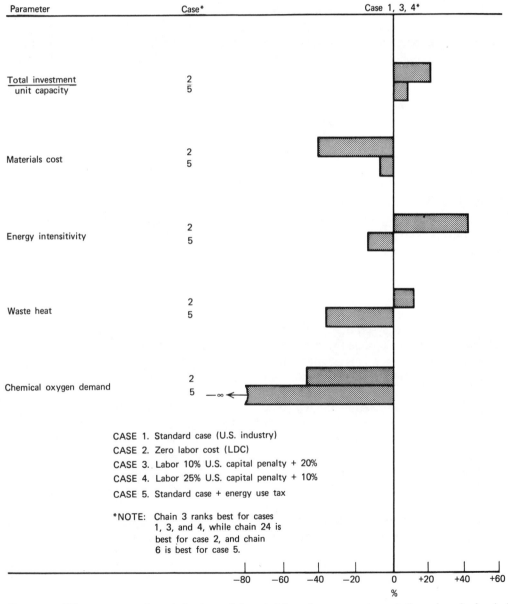

Figure 5.5 Percentage difference among best chains in capital, material, and energy costs, and thermal and chemical pollution (Source: Reference 36).

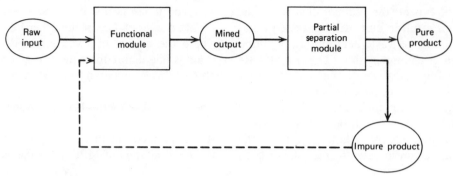

Figure 5.6. Schematic process loop.

physical inputs and outputs are fixed in relation to each other. This is a serious weakness for certain industries, however, where a variable mix of different products may be obtained from a given input or set of inputs, utilizing the same capital equipment in slightly different ways, at the option of a "manager."

A special but very important situation that frequently arises within materials-processing complexes —but rarely at a more aggregated level—is the "process loop" as illustrated in Figure 5.6 below. Typically a "raw" input is combined with a partially processed input. These are processed together and the resulting product then goes through a partial separation stage in which some pure product is obtained along with a significant quantity of impure product that must be recycled and processed further.

There are several possible versions of this loop, but the key element in Figure 5.6 is the fact that the impure or partially-processed by-product is quantitatively significant and *must* be recycled in a practical process. Hence the intermediate process steps must handle variable combinations of raw materials and partially processed materials, with the latter often predominating in importance, even though they never move outside the complex. The composition of the output stream is a function of the number of internal cycles, which is at the discretion of the manager, and hence a variable parameter as far as the modeller is concerned.

Examples of this kind of process loop are easy to identify. A distillation column is perhaps the classic example. Distillation is involved in many industrial processes, including the manufacture of grain spirits, petroleum products, and the extraction of salts from brines. Evidently, other types of partial separation devices, from centrifuges to filters, are also typically incorporated in loop arrangements.

From a model point of view, it is necessary to incorporate equations (or submodels) at each process node to define the relationship between "control" settings, or operating conditions, and input–output requirements. In effect, the problem is to introduce multidimensional production functions. To make the discussion concrete, let us consider the coal conversion sector—consisting of several existing processes plus a number of experimental or hypothetical possibilities—for converting coal into various finished products (high Btu gas, low Btu gas, "syncrude," methanol, or other chemicals). This sector is schematically outlined in Figure 5.7. Unlike the polyvinyl chloride–polyethylene network previously considered, however, *none of the intermediate products (except hydrogen) are marketable:* They are all produced and consumed within the same process-complex.

The conventions used in Figure 5.7 are similar to those used in Figure 5.4, except that circled materials are not necessarily marketable commodities and the squares represent "functional modules" (or subprocesses), rather than complete economic processes in the previous sense. We have increased the level of disaggregation of the analysis, which is analogous to increasing the magnification or "resolving power" of a microscope.[†]

It is important to note that intermediates may be in transient thermodynamic states, not at

---

[†]If an economic process can be regarded as a complete independent organism, its functional modules are comparable to organs of the body (heart, liver, digestive tract, etc.). Moreover, as with living organisms, it is possible to proceed via different routes involving different sets of functional modules, even though the basic functions (ingestion, elimination of wastes, defense, reproduction, adaptation to changing environmental conditions, etc.) are common to most or all living organisms.

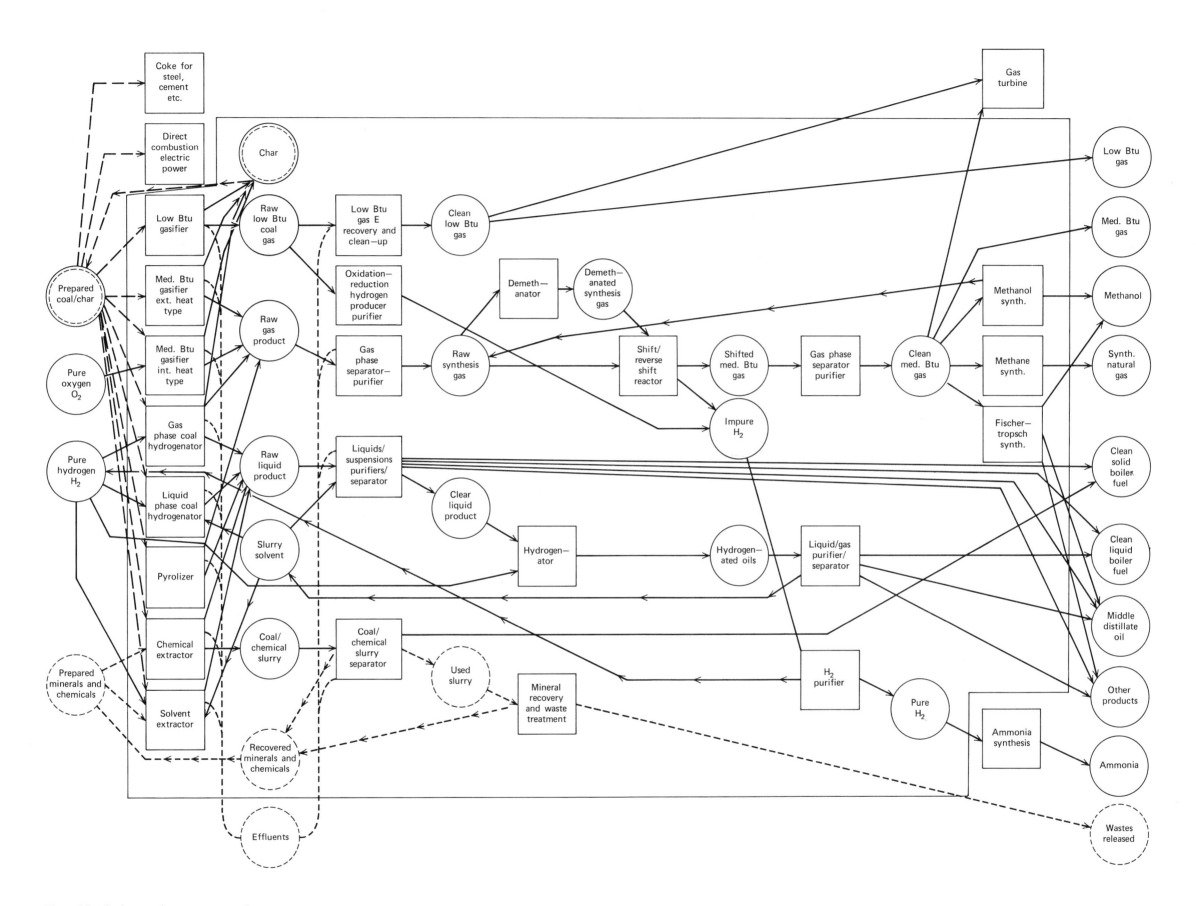

Figure 5.7. Coal conversion process network.

equilibrium with the ambient environment. For example, functional modules may operate at pressures or temperatures, which depend upon the chemical composition and the thermodynamic state of the input streams. Clearly, the performance of a functional module is likely to depend upon all these state variables. It is also important to note that most specific configurations for a functional module are applicable only to inputs that are provided in a specific *form* (e.g., slurry, liquid, and gas). These constraints are of utmost importance in practice.

As in the MPPM-I, a computerized PCEP is needed, but of a more sophisticated sort. A list of functional modules and competing configurations is provided with a dossier on each. The dossier specifies inputs, products, wastes, capital and labor requirements, as a function of each other, and of operating conditions. The problem of determining whether a chain is feasible is considerably more complex than in MPPM-I, however, since one cannot immediately determine whether a given module can be "linked" to another one or not. The general procedure is to start by specifying a final product and a raw material and initially identify all functional modules that will, in principle, produce it as a primary product, using a prespecified "road map" such as Figure 5.7.

This road map identifies all allowable links (via materials) between functional modules. It also identifies all allowable links, via functional modules, between different materials. The next step is to identify all competing *configurations* of equipment, the (range of) operating conditions necessary to yield the specified output, and the (range of) input requirements that are indicated for those operating conditions. A set of mathematical equations—in effect, a simplified model of the functional module—is obviously needed, and the PCEP must be able to "call in" from a random-access storage such a model whenever it is needed.

So far so good. But, is the link a feasible one? Unfortunately, the general procedure described here can sometimes yield a definite "no," but usually only a contingent "yes." The result of the analysis at this stage *may* have determined that some combination of operating conditions and inputs will yield the required product. In this case, the computerized search continues, link by link, until the sequence is broken by a definite "no" or the raw material stage is finally reached and a definite "yes" is justified. Meanwhile, the computer must store in its memory

all necessary information about each chain that remains in the running.

For modern computers, this is not as formidable a difficulty as might appear at first glance. Taking Figure 5.7 as an illustration, if there were eight alternative configurations for each functional module (a large number) and as many as eight modules on a path, the total number of possible chains would be $8^8 \gtrsim 10^7$. However, in practice, it would seldom be necessary to consider more than 10^3 possibilities, and usually far fewer. Nevertheless, the number of possible combinations could become unwieldy—and include an overwhelming amount of nonsense—unless the allowable conditions are strictly prescribed.

In the case of identifying a "feasible" sequence of configurations, the PCEP has already determined all relevant materials/energy requirements. This configuration is automatically scaled. Many configurations of equipment operate only for materials/energy throughputs within certain limits, so it can be ascertained immediately if there will be an overcapacity at some functional module because the smallest feasible module is bigger than necessary, or, conversely, if more than one unit will be needed because the largest-size single units available are inadequate. These considerations obviously bear on the economics of alternative chains. But, more important to the economic evaluation is the efficiency with which by-products can be recycled within the chain (i.e., within the plant) and the amount and type of waste treatment that will be required.

The comparison and rank ordering of chains of alternative configurations for functional modules in MPPM-II—as applied to the coal industry or any other—proceeds much as in MPPM-I. Any form of objective function, specifying raw material prices, interest rates, labor costs, environmental constraints, and so forth, can be utilized by the model.

REFERENCES

1. G. B. Dantzig, in T. C. Koopmans (ed.), *Activity Analysis of Production and Allocation*, Cowles Commission Monograph #13, Wiley, New York, 1951.

2. G. B. Dantzig, *Linear Programming and Extensions*, Princeton University Press, Princeton, NJ, 1963.

3. T. C. Koopmans, "Analysis of Production as an Efficient Combination of Activities," in T. C. Koopmans, op. cit. (Ref. 1).

4. L. V. Kantorovich, "On the Calculation of Production Inputs," translated in *Problems of Economics*, 1960.

5. R. Bellman, *Dynamic Programming*, Princeton University Press, Princeton, NJ, 1957.

6. L. S. Pontrjagin, V. G. Boltyanskii, R. V. Gamrelidze, and E. F. Mischchenko, *The Mathematical Theory of Optimal Processes*, K. N. Trirogoff (trans.), Wiley-Interscience, New York, 1962.

7. A. S. Manne, "A Linear Programming Model of the U.S. Petroleum Refining Industry," *Econometrica* (January 26, 1956).

8. H. Markowitz, "Industry-Wide Multi-Industry and Economy Wide Process Analysis," in T. C. Koopmans (ed.), *The Structural Interdependence of the Economy*, Wiley, New York, 1956.

9. W. Isard, E. W. Schooler, and T. Vietorisz, *Industrial Complex Analysis and Regional Development*, Harvard University Press, Cambridge, MA, 1959.

10. A. S. Manne and H. Markowitz (eds.), *Studies in Process Analysis*, Cowles Foundation Monograph #18, Wiley, New York, 1961.

11. K. Hoffman, "The United States Energy System—A Unified Planning Framework," Ph.D. Thesis, Brooklyn Polytechnic Institute, New York, 1972.

12. E. A. Cherniavsky, "Brookhaven Energy System Optimization Model," BNL 19569 for USAEC, December 1974.

13. "Reference Energy Systems and Resource Date for Use in the Assessment of Energy Technologies," AET-8, Brookhaven National Laboratories, Upton, NY, April 1972.

14. D. J. Behling, W. Marcuse, N. Swift, and R. G. Tessmer, Jr., "A Two-Level Iterative Model for Estimating Inter-fuel Substitution Effects," BNL 19863, Brookhaven National Laboratory, Upton, NY, March 1975.

15. C. S. Russell and W. O. Spofford, Jr., "A Quantitative Framework for Residuals Management Decisions," in A. V. Kneese and B. T. Bower (eds.), *Environmental Quality Analysis: Theory and Method in the Social Sciences*, Resources for the Future, Johns Hopkins University Press, Baltimore, 1972.

16. W. O. Spofford, Jr., C. S. Russell, and R. A. Kelly, "Operational Problems in Large Scale Residuals Management Models," in E. S. Mills (ed.), *Economic Analysis of Environmental Problems*. National Bureau of Economic Research, Columbia University Press, New York, 1975.

17. C. S. Russell, W. O. Spofford, Jr., and R. A. Kelly, "Interdependencies Among Gaseous, Liquid and Solid Residuals: The Case of the Lower Delaware Valley," *Northeast Reg. Sci. Rev.*, 5 (November 1975).

18. C. S. Russell, W. O. Spofford, Jr., and E. T. Haefele, "The Management of the Quality of the Environment," in J. Rothenberg and I. G. Heggie (eds.), *The Management of Water Quality and the Environment*, Halsted, New York, 1974.

19. A. V. Kneese, R. U. Ayres, and R. d'Arge, *Economics and the Environment: A Materials Balance Approach*, Resources for the Futures, Johns Hopkins University Press, Baltimore, 1970.

20. W. I. Zangwill, *Nonlinear Programming: A Unified Approach*, Prentice-Hall, Englewood Cliffs, NJ, 1969.

21. R. A. Kelly, "Conceptual Ecological Model of the Delaware Estuary," in *Systems Analysis and Simulation in Ecology*, Vol IV, Academic Press, New York, 1974.

22. H. W. Streeter and E. B. Phelps, "A Study of the Pollution and Natural Purification of the Ohio River," Public Health Bulletin No. 146, U.S. Public Health Service, Washington, DC, 1925.

23. C. O. G. Löf and A. V. Kneese, *The Economics of Water Utilization in the Sugar Beet Industry*, Resources for the Future, Washington DC, 1968.

24. C. S. Russell, *Residuals Management in Industry: A Case Study of Petroleum Refining*, Johns Hopkins Press, Baltimore, 1973.

25. C. S. Russell and W. J. Vaughan, "A Linear Programming Model of Residuals Management for Integrated Iron and Steel Production," *J. Environ. Econ. Management*, 1 (1) (May 1974).

26. B. T. Bower, C. O. G. Löf, and W. M. Hearson, "Residuals Management in the Pulp and Paper Industry," *Nat. Resources J.*, 11 (4) (1971).

27. R. G. Thompson, A. K. Schwartz, Jr., and D. Y. Slimak, "An Industrial Economic Model of Water Use and Waste Treatment for a Representative Ethylene Plant," University of Houston, mimeo, April 1973.

29. R. G. Thompson, D. Y. Slimak, A. K. Schwartz, Jr., and K. Dew, "A National Economic Model of Water Use and Waste Treatment for Olefins Production," University of Houston, mimeo, 1974.

30. J. A. Calloway, A. K. Schwartz, Jr., and R. G. Thompson, "Industrial Economic Model of Water Use and Waste Treatment for Ammonia," University of Houston, mimeo, 1974.

31. R. Thoss, "A Generalized Input–Output Model for Residuals Management" in K. R. Polenske and J. Skolka (eds.), *Input–Output Techniques*, Ballinger Press, Cambridge, MA, 1976.

32. R. Thoss and K. Wiik, "Optimal Allocation of Economic Activities Under Environmental Constraints in the Frankfurt Metropolitan Area," paper prepared for the International Econometric Association Conference on Econometric Contributions to Public Policy, Urbino, Italy, September 2–8, 1976.

33. A. S. Manne, "Waiting for the Breeder," *Rev. Econ. Stud.: Symposium on the Economics of Exhaustible Resources* (1974).

34. L. E. Westphal, *Planning Investments with Economies of Scale*, North-Holland, Amsterdam, 1971.

35. L. E. Westphal, "Planning With Economies of Scale," in C. R. Blitzer, P. B. Clark, and L. Taylor (eds.), *Economy-Wide Models and Development Planning*, for the World Bank, Oxford University Press, Oxford, 1975.

36. A. S. Manne, "A Mixed Integer Algorithm for Project Evaluation" in L. M. Goreux and A. S. Manne (eds.), *Multi-Level Planning: Case Studies in Mexico*, North-Holland, Amsterdam, 1973.

37. L. E. Westphal and Y. W. Rhee, "The Allocative Consequences of Economies of Scale," presented at Econometric Society Session, ASSA Meetings, 1974.

38. R. U. Ayres, "A Materials–Process–Product Model" in A. V. Kneese and B. T. Bower (eds.), *Environmental Quality Analysis: Theory and Method in the Social Sciences*, Resources for the Future, Johns Hopkins University Press, Baltimore, 1972.

39. R. U. Ayres and M. O. Stern, "An Activity Analysis Approach to Project Feasibility Analysis for Industrial Development" (For UNIDO), IRT-296-P, Washington, DC, September 1972.

40. R. U. Ayres, J. Cummings-Saxton, and M. O. Stern, *Materials Process Product Model: A Feasibility Demonstration Based on the Bottle Manufacturing Industry*, for the National Science Foundation under Contract NSF-C-652, IRT-305-FR, Arlington, VA, June 1974.

41. R. W. Roig, A. J. Jagota, P. J. Zammato, N. E. Leggett, and D. S. Soni, "Materials-Process-Product Analysis of Coal Process Technology," Final Report, IRT-494-FR, under ERDA Contract E(49-18)-2027, October 10, 1977.

42. E. Kirk and D. F. Othmer, *Encyclopedia of Chemical Technology*, Vols. 1–25, Wiley, New York, 1963–1972.

43. M. S. Peters and K. D. Timmerhaus, *Plant Design and Economies for Chemical Engineers*, 2nd ed., McGraw-Hill, New York, 1968.

44. F. A. Lowenheim and M. K. Moran, *Industrial Chemicals*, 4th ed., Wiley, New York, 1975.

45. K. M. Guthrie, "Capital and Operating Costs for 54 Chemical Processes," *Chem. Eng.*, 140–156 (June 15, 1970).

46. J. T. Gallagher, in Herbert Popper (ed.), *Modern Cost Engineering Techniques*, McGraw-Hill, New York, 1970, pp. 3–10.

47. H. J. Lang, in C. H. Chilton (ed.), *Cost Engineering in the Process Industries*, McGraw-Hill, New York, 1960, pp. 3–13.

48. G. Enyedy, "Cost Data for Major Equipment," *Chem. Eng. Process*, **67** (5), 73–81 (1971).

49. F. C. Zevnik and R. L. Buchanan, *Chem. Eng. Process*, 70, (February 1963).

APPENDIX 5-A

COST ANNUALIZATION

Methods for determining the annual costs of investment and of operation differ. Operating costs are relatively evenly distributed over time and can be directly computed on an annual basis. Fixed costs, however, are incurred at the outset and at periodic intervals in the venture. These costs must be "annualized" in some manner. Several annualization schemes are available; they differ mainly in their treatment of depreciation and amortization of investment.

It is convenient to divide capital requirements into fixed and working capital, as briefly discussed in Section 5.3. Fixed investment includes "battery limits" plant, supporting equipment, and land.

Battery limits plant refers to the assemblage of tanks, pipes, towers, pumps, and the like, required to manufacture the product. This investment is relatively independent of whether an addition to an existing facility or an entirely new plant is being considered. Supporting, or offsite, investment is the fixed investment in everything else needed for the operating facility. It includes auxiliaries, materials storage and handling, and services. Auxiliaries include fuel, power, steam, refrigerants, water and sewer, and buildings. Materials handling and storage refers to facilities outside the battery limits plant (except in-process storage). Services include lighting, telephone, shops, washrooms, lockers, and so on. Land the working capital are considered separately from other capital requirements. A comprehensive checklist of the items to be considered under each heading may be found in standard references, such as the *Chemical Engineers' Handbook*.[A1]

To annualize the investment costs, consider an investment I in a new venture (such as a unit operation in a new process chain) made at the beginning of the project, that is, at $t = 0$. We subdivide investment into three parts, as follows:

$$I = I_F + I_A + I_W \qquad (i)$$

where I represents the total investment in the unit operation, I_F refers to fixed installations attributable to the venture (i.e., "battery limits" plant), I_A is for auxiliary supporting facilities, and I_W is for working capital and land. This subdivision is convenient because it groups investments according to their useful life, which in turn determines their rate of depreciation.

We start with a simple assumption, namely that the new venture is financed entirely by money borrowed at an interest rate i comparable to what the money could earn if invested elsewhere, and that the investment I has unlimited life expectancy. That is, like land, it does not depreciate. The problem is to calculate equal annual costs to pay for this investment. These costs must equal the minimum annual earnings required to justify the investment. Obviously, under these circumstances, the annual payments are iI, since the investment does not depreciate, hence does not have to be amortized.

Now, suppose a finite venture life of m years for the plant. We demand that the present worth of the venture W, when discounted at rate i (which the investment could earn without depreciating if placed elsewhere), should be at least equal to the investment I:

$$W = \frac{iI}{1+i} + \frac{iI}{(1+i)^2} + \frac{iI}{(1+i)^3} + \cdots$$

$$+ \frac{iI}{(1+i)^m} + \frac{I}{(1+i)^m}$$

$$= iI\left[\sum_{n=1}^{m} \frac{1}{(1+i)^n}\right] + \frac{I}{(1+i)^m}$$

$$= iI\left[\frac{1}{1-1/(1+i)} - 1\right]\left[1 - \frac{1}{(1+i)^m}\right]$$

$$+ \frac{I}{(1+i)^m} \equiv I \tag{ii}$$

This result is obtained by adding earnings at the end of the first year, iI, discounted to time $t = 0$ by the factor $(1 + i)$, those at the end of the second year, also iI, but discounted to the present by the factor $(1 + i)^2$, and so on, up to the end of the mth year, when the last interest payment iI is due, discounted to present value as $iI/(1 + i)^m$. At that time, the venture terminates, and the original investment I is recovered intact. Since it is undepreciated, it contributes an amount $I/(1 + i)^m$ to the present worth of the venture. The series is summed in closed form with the help of the relation

$$1 + x + x^2 + \cdots = \sum_{n=0}^{\infty} x^n = (1 - x)^{-1}, \text{ for } x < 1,$$

where, in our case, $x = (1 + i)^{-1}$. Thus I is obtained as the present worth of the venture, no matter what m is used, provided the discount rate is i and the annual earnings are iI. If we let $m \to \infty$, the present worth of the recovered investment disappears, still leaving I.

The notion of using borrowed money at prevailing interest rates for a risky venture is not realistic in most instances. Prevailing interest rates are bolstered predominantly by senior securities issued by ventures with a long record of stable operation. Most companies raise their initial capital as well as some subsequent financing by issuing equity; and subsequent ventures may be financed in part by equity, by accumulated surplus (from past profits), and/or by debt. The ratio of debt-to-equity financing varies from industry to industry, and from company to company within each industry. If we are not greatly interested in such details, but merely wish to compare and account for the *ceteris paribus* economics of unit operations, unit processes, and process chains under varying constraints, we can assume that a given company can be characterized by an internal interest rate i that plays the same role as the previously assumed external borrowing rate, but is generally higher because it allows a typical rate of return to the stockholders for their assumption of risk, as well as an interest rate on use of the capital funds I, adjusted for tax allowances.

Thus the rate i represents the marginal net return that would be obtained by investing I in existing company ventures. This is the "minimum acceptable rate of return on investment" that the new venture would have to earn to compete successfully for I. It represents an opportunity cost for the new venture and determines the rate at which future venture earnings should be discounted to determine present worth. If the new venture's risk is higher than that for the company as a whole, as it might be for an innovative process, the discount rate for that process should be raised to $i + h$, $h \geqslant 0$, and the expected yearly net return from the investment I in this venture should be $(i + h)I$. In short, the fact that a company is willing to make an investment I in a new venture at return i indicates that it sees opportunities for investment in new ventures in its industry at that rate, and that it has been willing to set aside capital for such an investment, rather than putting it internally to work in established ventures.

We must now take into account the fact that part of the investment, namely I_F and I_A, does not have indefinite life, but depreciates with time through usage (or obsolescence). At least two schemes (with variations) are available for dealing with this added complexity, depending on how i is defined. One might define i as the minimum acceptable rate of return for tying up an investment in a nondestructible, nondepreciable asset such as working capital or land. In that case, a higher rate of return than i should be demanded of a depreciable asset. This can be done, for example, by establishing a sinking fund for replacing the asset when its value terminates. This is the method used by Rudd and Watson.[A2] Alternatively, we might define i as the minimum acceptable rate of return on an investment in an asset having a life comparable to that of the venture we are considering. In that case, were it not for taxes, depreciation need not be taken into account. Any asset with remaining value (such as I_W) at the termination of the venture would be discounted to the present and credited against the venture worth, as done in Equation (ii). We adopt the first method, because it permits a more reliable and stable initial determination of i. However, the two methods are equivalent with suitable choices of i, and each has its adherents.

In using the sinking fund method, an amount eI is set aside at the end of each of m years equal to the life of a venture, and the accumulating sinking fund is invested each year in other company ventures with rate of return i, so that at the end of m years the accumulated assets equal I. We assume, for simplicity, that venture life m and the lives of investments I_F and I_A are identical. At the end of m years, the sinking fund is

$$eI\left[1 + (1 + i) + (1 + i)^2 + \cdots + (1 + i)^{m-1}\right]$$

$$= \frac{eI}{i}\left[(1 + i)^m - 1\right]$$

We require that this amount equal initial investment,

$$I = \frac{eI}{i}[(1 + i)^m - 1]$$

Then   (iii)

$$e = \frac{i}{(1 + i)^m - 1}$$

The annualized capital cost C_I of an investment I, to be amortized over a period m, is, therefore,

$$C_I = (i + e)I = iI\left[1 + \frac{1}{(1 + i)^m - 1}\right]. \quad \text{(iv)}$$

The equation used in PCEM is somewhat more general than Equation (iv):

$$C_I = (i + h)\,I\,(1 - Df_D)$$
$$+ i\left(\frac{I_F + I_A}{(1 + i)^m - 1}\right) + jDIe \quad \text{(v)}$$

where h is the high risk rate of return, and DI is the portion of the investment cost I raised through borrowing at rate j. The parameter f_D can take on values between 0 and 1, depending on industry practice. If f_D is set equal to zero, the rate of return on the full investment is $(i + h)$, even though part of the capital is borrowed; if $f_D = 1$, the annualized capital cost allocates a return $(i + h)$ only on that part of the investment committing the company's own equity. There are several other parameters used in our computerized expression to permit greater flexibility. The principle remains the same.

Equation (v) represents the annualized "investment cost" of the venture. Without the term jDI, treated as an operating cost, Equation (v) corresponds to the minimum acceptable net annual return on investment I, which we denote as C_I'. We assume that C_I' represents the desired (or target) net annual return or net profit. We calculate what the gross profit must be to yield the target net profit by adding terms corresponding to various types of taxes paid or subsidies received. This allows us to develop an expression for the total value added, with all components on an annual basis. We do not attempt to present a fully general expression, since taxes and subsidies vary greatly from country to country in application and method of calculation. However, it is worthwhile generalizing Equation (v) to include several specific forms of tax, including income tax, investment credit and value-added tax.

REFERENCES

A1. R. H. Perry, C. H. Chilton, and S. P. Kirkpatrick, *Chemical Engineers' Handbook*, 4th ed., McGraw-Hill, New York, 1963, pp. 26-1–26-45.

A2. Rudd and Watson, *Strategy of Process Engineering*, Wiley, New York, 1968.

CHAPTER SIX *The Need for an Integrated Materials/Energy Balance Statistical System (MEBSS)*

This book has already discussed the need for a revised and extended "paradigm" for economics to reflect expansion into fields that were formerly ignored or treated as slightly disreputable—if not illegitimate—relatives of the "main branch" of the economics family, by which I mean the study of production, consumption, and capital formation. In particular, I have attempted to integrate resource and environmental aspects of economics in a fashion that is cognizant of, and consistent with, the basic laws of physical science. Admittedly, the treatment is far from complete: The focus of the book is on realistic, large-scale simulation or optimization models of potential interest to policy-makers. Many topics in the vast literature of resource or environmental economics obviously have been left untouched. One must draw lines somewhere.

However, I cannot conclude a book about models without a discussion of their statistical underpinnings, since many of the practical limitations of large-scale resource/environmental economic models are attributable to data problems. As Wassily Leontief, the father of input–output modeling said in his Presidential address to the American Economic Association in 1971,[1]

The shift from casual empiricism that dominates much of today's econometric work to systematic large-scale factual analysis will not be easy . . . . The spectacular advances in computer technology increased the economists' potential ability to make effective analytical use of large sets of detailed data . . . . What is, however, urgently needed is the establishment, maintenance and enforcement of coordinated uniform classification systems . . . . Incompatible data are useless data . . . . How far from a tolerable, not to say, ideal state our present economic statistics are in this respect, can be judged by the fact that because of differences in classification, domestic output data cannot be compared, for many goods, with the corresponding export and import figures. Neither can the official employment statistics be related without laborious adjustments to output data, industry by industry . . . . An unreasonably high proportion of material and intellectual resource devoted to statistical work is now spent not on the collection of primary information but on a frustrating and wasteful struggle with incongruous definitions and irreconcilable classifications.

By the same token, it is entirely realistic to assume that the growing demand for such models will stimulate the development of more appropriate statistical series. The situation here is entirely analogous to the development of national account statistics several decades ago in response to a growing interest in short-run business cycle and long-run input–output models. It need hardly be added that as the available statistics were improved the models performed better, and thus gradually stimulated still greater demand for such models—and for statistics.

It is hardly surprising that the present state of environmental economic models is not as advanced as that of business cycle or "growth" models in view of the data deficiencies that currently exist. In Chapters 4 and 5, I have focused the discussion on the formal aspects of model structure. Yet the sophisticated reader must be aware that implementation of materials/energy or resource/residual simulation or optimization models requires very extensive physical stock and flow accounting data inputs, far beyond what is required for their "core" economic models.

The next several sections of this chapter describe a proposed materials/energy balance statistical system (MEBSS) to fill the needs of resource/environmental policy models.[†] It must be emphasized that the scheme described in the remainder of this chapter is ambitious and full of difficulties. These difficulties may, at any point, dictate modifications or simplifications. At best, practical realization of the proposed system—even in partial form—is probably a decade, or more, in the future.

6.1 BASIC STRUCTURE AND DESIGN OF MEBSS

The underlying design principle for the proposed Materials/Energy Balance Statistical Subsystem is conservation of materials and energy: All material and energy inputs to the world economic system, as well as to individual countries, must be accounted for either as final outputs or as changes in ac-

cumulated stocks, including durable goods in service as well as inventories. Two "balance" concepts are used: A gross (volume) balance is applied in the case of production, consumption, and trade of major resources and commodities. A more refined materials and energy balance, by process, is also applied to elucidate the relation between resource/commodity production and consumption and the generation of waste flows.

The terms "stock" and "flow" are used in the foregoing paragraph in a general and somewhat undefined sense. Definitions of these terms, as specifically used in the design of MEBSS, are discussed later. Distinctions must be made between stocks that are physically distinguishable natural reservoirs, such as groundwater or oceans, and stocks that are essentially accounting categories, differentiated by ownership, production processes, and the like. Similarly, flows may be additions or subtractions to the quantity of a physical resource contained in a given natural reservoir, or they may be transformations from one physical form to another, or they may represent shifts from one accounting category to another.

The fact remains that all changes of stocks are equivalent to flows, and this accounting identity is essential in constructing a statistical system. It follows, too, that changes in a stock can either be determined by direct enumeration and/or direct measure, or they can be inferred by measuring and summing the corresponding input and output flows. Where both possibilities are feasible, one provides a useful check on the other. Sometimes only one of the two approaches is feasible, however, and MEBSS need not specify which approach is to be used in any given case.

Because of the "interconnectedness" of the system of materials/energy stocks and flows, it is natural to employ a diagrammatic form of presentation. MEBSS is displayed graphically in Figure 6.1 and in matrix form in Table 6.1. The numbered blocks in Figure 6.1 correspond to major accounting categories.

The system can best be explained in terms of an economy utilizing only one basic material resource (say, wood), which is extracted from the environment. The "opening stock" in year t is the total known reserve of wood available (e.g., in the forest) at the beginning of the year. The amount of wood actually extracted from the forest during the year is $T_{1.2}$.

[†]The material presented here is based largely on a working paper prepared by the author for the United Nations Statistical Office.[2] The concept has been substantially incorporated in the Report by the Secretary General to the UN Statistical Commission (November 1976 Meeting in New Delhi) recommending further development of MEBSS by the UN Statistical Office. However, I must make it clear that the Statistical Commission endorsement is both limited and contingent. Neither feasibility nor acceptability to potential users is estimated as yet. Ultimate adoption in anything like the form discussed here is highly uncertain. Needless to say, in recommending the scheme, I speak for myself only.

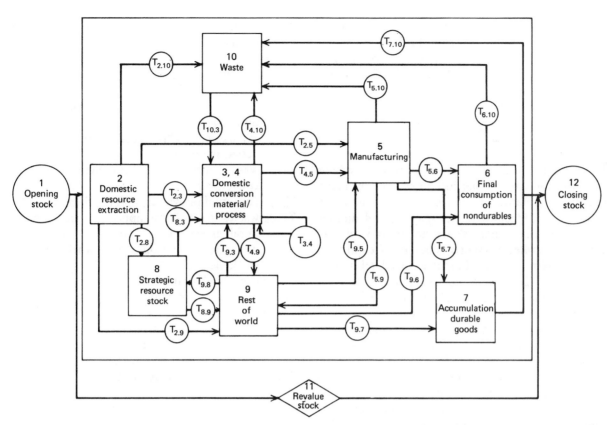

Figure 6.1. Systems of environmental statistics (diagrammatic form).

Table 6.1 Matrix Form of System of Environmental Statistics

		1	2	3	4	5	6	7	8	9	10	11	12
1	Opening stock												
2	Domestic extraction		$T_{1.2}$	$T_{2.3}$	$T_{2.4}$	$T_{2.5}$			$T_{2.8}$	$T_{2.9}$	$T_{2.10}$		
3	Domestic conversion (proc.)				$T_{3.4}$						$T_{3.10}$		
4	Domestic conversion (mat.)					$T_{4.5}$				$T_{4.9}$			
5	Intermediate production						$T_{5.6}$	$T_{5.7}$		$T_{5.9}$	$T_{5.10}$		
6	Final consumption of nondurables										$T_{6.10}$		
7	Accumulation of durable goods										$T_{7.10}$		
8	Strategic resource stock			$T_{8.3}$						$T_{8.9}$			
9	Rest-of-world			$T_{9.3}$		$T_{9.5}$	$T_{9.6}$	$T_{9.7}$	$T_{9.8}$				
10	Waste		$T_{10.3}$										
11	Revaluation of stock												
12	Closing stock												

173

The amount of wood "converted" to other forms (say, lumber, plywood, and paper) would be a vector of transformation flows $T_{2.3}^{(i)}$ if classified according to the conversion process used, and $T_{2.4}^{(j)}$ if classified according to the form of the product. Evidently, the totals must be the same:

$$\sum_i T_{2.3}^{(i)} = \sum_j T_{2.4}^{(j)}$$

Intermediate production refers to wood products produced for sale to other industries (e.g., furniture, housing). Domestic final consumption of non-durable wood products (e.g., newspapers) and durable goods (e.g., chairs, houses) are straightforward concepts. The strategic resource stock is, of course, an inventory term that may fluctuate considerably from year to year. Trade with the rest of the world comprises imports and exports at any stage of processing. Wasteflow refers to losses occurring anywhere in the economy, from extraction to consumption. "Revaluation of stock" is the change in known reserves during the year, which arises either from (1) natural gains (e.g., growth) or losses (e.g., from forest fires), (2) improved extraction or utilization technology (e.g., development of methods to utilize stumps and branches), or (3) increased recoverability due to higher prices.

Based on this picture it is clear that definite relationships exist between the accounting categories, which can be expressed in mathematical terminology. To introduce additional resources causes no fundamental difficulty; it merely adds complexity. The transaction terms can be written down immediately:

$T_{1.2}$ = extraction of raw materials (energy) resources from available reserves (opening stock)

$T_{2.3}$ = conversion of raw materials to other forms, classified by converted material (j)

$T_{2.4}$ = conversion of raw materials to other forms, classified by conversion process (i)

$T_{2.5}$ = use of raw materials without conversion in intermediate production

$T_{2.8}$ = additions to strategic stockpile of raw materials

$T_{2.9}$ = exports of raw materials

$T_{2.10}$ = wasteflows arising from the extraction stage

$T_{3.4}$ = allocation of conversion processes among converted materials

$T_{3.10}$ = wasteflows from conversion, classified by process

$T_{4.5}$ = outputs of converted materials to intermediate production

$T_{4.9}$ = exports of converted materials

$T_{5.6}$ = allocation of intermediate production to final non-durables

$T_{5.7}$ = allocation of intermediate production to final durables

$T_{5.9}$ = export of industrial products other than converted materials

$T_{5.10}$ = wastes from industrial production other than conversion

$T_{6.10}$ = consumption wastes, non-durables

$T_{7.10}$ = wastes from consumption (depreciation) of durables and capital goods

$T_{8.3}$ = inputs to conversion processes from strategic stockpile

$T_{8.9}$ = exports of strategic stocks

$T_{9.3}$ = imports of raw materials for conversion

$T_{9.5}$ = imports of converted materials for intermediate production

$T_{9.6}$ = imports of final goods, non-durable

$T_{9.7}$ = imports of final goods, durable

$T_{9.8}$ = imports of raw materials for the strategic stockpile

$T_{10.3}$ = reconversion (recycling) of waste materials

It is noted that balances can be defined at several different levels of aggregation. In the first place, all material and energy input and output flows to a given category or "sector" in Figure 6.1 must exactly be accounted for by shift to another category: import, export, conversion to another form, use or embodiment as such in a manufactured product, or depletion or accumulation of some stock. This must hold not only for aggregate quantities (measured in mass or weight units, for instance) but also for inputs and outputs of specific chemical elements. On the other hand, each chemical element (e.g., sulfur or chlorine) must be accounted for as a whole; that is, outputs of an element by one sector or activity must be balanced by inputs to others, and conversely. For instance, all coal domestically produced (plus imports less exports) must be accounted for exactly by the various uses (conversion to coke, combustion as such) or by changes in inventory. By extension, of course, the same kinds of balances must also hold for groups of sectors, or for the economy as a whole.

The foregoing paragraphs may leave a deceptive impression of simplicity. The inherent difficulties begin to be apparent when it is recognized that

MEBSS must trace transformations of materials and energy from one form (or category) to another, starting with raw materials and ending with waste products. For instance, coal is used to generate electricity, another form of energy. Clearly, MEBSS cannot treat electricity generated from coal-fired boilers as an item in the accounts equivalent to coal itself, even though some users might consider them as mutually substitutable. From the statistical viewpoint, processed forms of materials or energy should be kept separate from raw forms. Moreover, each successive stage of processing must be explicitly distinguished.

Another difficulty is posed by the interconvertability of fuels and materials for other purposes. Natural gas, oil, coal, and wood can be burned for heat. Or coal can be converted to coke, which is used to reduce iron ore. The chemical energy value contained in the coke is partly wasted in the form of heat and by-products, such as CO_2, and partly embodied in the product (pig iron). Note that if the pure iron were oxidized (i.e., burned), it would revert back to the chemical form of the ore (Fe_2O_3 or Fe_3O_4). Thus reduction is, energetically, the reverse of oxidation. It follows, of course, that pig iron or ingot aluminum "embody" a certain amount of energy. This energy must appear somewhere in the accounts.

An example that is currently quantitatively insignificant, but may be important in the future is the use of refined metals such as lead, zinc, and nickel as battery anodes. (In the future, lithium, sodium, or magnesium may be widely used in this application.) As the battery is discharged, of course, the metal in the anode reacts with oxygen or some other element, resulting in a lower "available" energy content. The process may be reversed by "recharging" the battery using an external electrical supply. A satisfactory statistical system must be capable of disentangling these energy conversion processes.

Another troublesome problem arises when natural gas, together with atmospheric air, are converted into ammonia (NH_3). The hydrogen in the ammonia comes originally from the methane in the gas (CH_4). Thus the gas is, in effect, a feedstock for a nonfuel chemical product. (Of course, the ammonia has energy content and could actually be used as a fuel.) Natural gas is also the main source of ethane and propane, which are in turn converted to ethylene (C_2H_4) and propylene (C_2H_6), the basic building blocks of organic chemistry, leading to synthetic fibers, plastics, and so forth. In the same way, butane, benzene, toluene, and xylene are generally obtained from cracking crude oil, or (less frequently) from coal tar. These are used, in turn, to manufacture butylene, butadiene, and styrene and a host of other synthetic elastomers and polymers.

Obviously, some materials that are manufactured from potential primary fuels later become available again as "secondary" fuels. This applies especially to organic materials, such as paper and paperboard. Similarly, secondary flows of nonfuel materials such as scrap iron and waste paper must be considered as sources of energy to the corresponding materials processing sectors. A satisfactory accounting system should explicitly reflect these factors.

It is important to note, also, that many processed materials can be derived from alternative—sometimes very diverse—sources. Acetic acid and methyl alcohol can be obtained commercially from destructive distillation (pyrolysis) of wood or from fossil hydrocarbons. Ethyl alcohol for human consumption is derived from distillation of fermented carbohydrates (grain, potatos, fruit, etc.), whereas industrial alcohol is produced—much more cheaply—from natural gas liquids. Benzene and toluene are industrially produced from coal tar or from petroleum. Sodium carbonate (soda ash) is currently obtained mainly from natural deposits (trona), whereas it was formerly synthesized by the Solvay process from sodium chloride (rock salt) and limestone. The reverse is true of aluminum sodium fluoride (cryolite) used as a flux in the aluminum industry, which was formerly obtained from a natural deposit in Greenland but is now almost all manufactured synthetically. (In the future its use may be obviated by a process change).

Many industrial chemicals that are now produced directly from raw materials will be derived largely or entirely from waste products in the future. The most obvious example is sulfuric acid, which is currently manufactured from elemental sulfur but will eventually be a by-product of coal or oil desulfurization or stack-gas treatment. Examples are not confined to the chemical and metallurgical industries. For instance, plaster-of-paris—used in the building industry—is currently manufactured from the natural mineral gypsum, but will probably be available as a by-product of future stack-gas treatment processes. Future insulation, paving materials, and other construction materials will undoubtedly be derived increasingly from recycled wastes, and already many food products and additives are being derived from nonagricultural sources. This is notably true of synthetic beverages and sweeteners. But animal feedstuffs, and even synthetic proteins, may be produced in important quantities from fossil hydrocarbons before the end of this century.

An important consequence of the foregoing, from the standpoint of designing a set of tables, is that groupings that might seem "natural" today may not be so a few years hence. In particular, it *cannot* be assumed that a given converted material always comes from the same raw material or that a given raw material or ore will always be refined into the same final product. Thus while it is sometimes useful to distinguish raw (unprocessed) materials from converted materials, this cannot be done meaningfully within a more narrowly defined category such as "agricultural products," "fuels," or "chemical products."

The interconversion of materials and forms of energy raises a potentially troublesome set of problems pertaining to units of account. Various inputs to a physical or chemical conversion process may be traditionally measured in different units; similarly, inputs and outputs may be measured in different units. For instance, coal entering a generating plant is normally measured in mass (tonnage) units, whereas its energy content is typically expressed in Btu. For consistency, it is preferable to use the metric units, kilojoules (kJ). Electricity output is normally measured in kilowatt-hours in the United States. For international purposes, however, "electric" joules (kJe) will certainly be adopted. The equivalence factor between a given mass (e.g., a metric ton) of coal and the energy content of electricity generated from it is not a natural constant: it depends on the quality of the coal and the efficiency of the generating plant, which varies over time. The "energy balance" between inputs and outputs also involves different units (e.g. Btu in versus kWh out or kJ in versus kJe out).

A similar situation arises in the case of petroleum refineries where simultaneous materials and energy balances are needed. The input crude oil is normally expressed in terms of physical volume (usually barrels or bbl) of a certain specific gravity; each type of crude oil is characterized by a certain average heat energy content (Btu/bbl). For international purposes, common metric measures must be adopted (e.g., cubic meters, metric tons, and kilojoules per cubic meter). The outputs of refineries are also typically measured in volume terms, but the energy per unit volume of various refinery fractions (gasoline, naphtha, fuel oil, residual oil) varies quite considerably due to differences in specific gravity. *Thus the total volume of outputs need not be exactly equal to the volume of inputs.* But the sum of weights and energies should, of course, balance when refinery losses are considered. The specific gravity of input crude oil[†] will vary from one country to another; similarly, the specifications (including specific gravity) of outputs will also vary. Thus a set of supplementary weight/volume/energy equivalences must be supplied for *each* conversion process and for each country. Many of the indicated equivalence tables will, of course, be required for international energy statistics.

It is important to bear in mind that, because of its focus on materials/energy balances, the MEBSS will *not* be useful for cataloging all actual (or potential) pollutants. Specifically, MEBSS will necessarily concern itself with volume pollutants (such as organic wastes or combustion products), plus specific major chemicals and metals. However, some pollutants are not subject to conservation rules. These include noise—a nonconserved form of waste energy—"visual pollution" and litter, carcinogenetic, mutagenetic, teratogenetic, or toxic organic chemicals, and radioactive wastes. These can only be dealt with by linking MEBSS to supplementary tabulations, such as the proposed IRTC.[‡]

Logically, MEBSS could very well be extended to include the "stocks and flows" of some elements such as carbon, nitrogen, phosphorus (and energy) in major environmental reservoirs, such as the atmosphere, soil, forests, surface waters, and oceans. At present, this must be regarded as an objective for the distant future, since it is not practically feasible to quantify most of the flows at present. From an environmental point of view, this gap is a major challenge that must be filled, in the first instance, by design of appropriate environmental monitoring systems.[§]

[†]In tropical countries, crude is often "spiked" by addition of naphtha to increase the output of light fractions, since heavy fuel fractions (used for heating) are less required.

[‡]*I*nternational *R*egistry of *T*oxic *C*hemicals. The linkage would be essentially automatic as long as both systems use the same (or related) classifications.

[§]For instance, the *G*lobal *E*nvironmental *M*onitoring *S*ystem (GEMS) as currently being developed by UNEP.

Similarly, ecological and health effects associated with pollution cannot now be incorporated within MEBSS. Such statistics are currently compiled to a minor extent by national agencies, FAO and WHO. Similarly, existing industrial and resource statistics must also be coordinated. Links with MEBSS can be provided by developing supplementary tables to reconcile relevant definitions and categories. At some future time, further modules for the System of Environmental Statistics will presumably be developed to incorporate these external data on a uniform and consistent basis.

6.2 TAXONOMIC CONSIDERATIONS

As noted already, the data base for the "core" economic models is largely derived from national account statistics.[†] An internationally standardized version, the United Nations System of National Accounts (SNA) has been evolving for the last three decades.[3] Not all countries adhere exactly to the UN standard, but its existence has certainly tended to encourage convergence toward a common set of definitions and taxonomy.

It seems clearly desirable that any international system of materials/energy balance account statistics should be designed explicitly to be compatible with the international SNA, notwithstanding its well-known deficiencies (and many errors in the figures, some deliberately introduced by participating countries). It should have a corresponding structure and all industry and commodity categories used in the former should be consistent as far as possible with the latter, even if greater disaggregation is required. This means among other things, that the standard *production/consumption* dichotomy[‡] used in the SNA should be retained in MEBSS. If it is ultimately modified in SNA it should, of course, be modified also for MEBSS. Finally, the International Standard Industrial Classification (ISIC)[4] used in SNA and the related International Standard Classification of Goods and Services (ICGS)[5] should be used for MEBSS.

Without going into excessive detail, the ICGS is an eight-digit code of which the first four digits correspond to the ISIC sector in which the commodity or service is produced. The next two digits identify classes of commodities or services, and the last two digits identify specific materials or closely related subclasses. As in any classification system, there are a number of "basket" categories for minor activities or products not explicitly classified (nec). Changes in relative importance of activities or commodities result in a need for periodic revision of the classifications—usually by successively breaking out more items from the nec groups. For MEBSS, the ICGS code will probably have to be extended to include some nontraded resources.

Incidentally, it is noteworthy that two other international systems of classification exist, namely, the Standard International Trade Classification (SITC)[6] and the Brussels Trade Nomenclature (BTN)[7] Code. Each of the defining documents contains a detailed cross-classification, so that it is easy in principle to determine all three codes for any given commodity. However, the fact that three different classifications are constantly being revised greatly complicates the problems of statistical record keeping.

Unfortunately, it is not possible to use the same concepts and definitions of capital goods, investment, accumulation, and depreciation (capital consumption) in both systems. This is because MEBSS must account for all forms of material accumulation in the same way. This means that durable goods normally counted as "consumption"—that is, housing, automobiles, appliances, and the like—will be lumped together, instead, with "capital goods," which are also durable in the same sense. That is, they do not contribute to waste streams until the lapse of a number of years (or decades).

Another taxonomic problem area arises with respect to resource statistics. Here there is no agreed international terminology as regards resource categories (e.g., "proved–probable–possible" and/or "measured–indicated–inferred" reserves).[8] In recognition of these ambiguities, the UN Economic and Social Council passed a resolution at its 1975 plenary meeting "recognizing the need to find agreement on terminology used on categorizing mineral resources so that there should be comparable and generally agreed statistics." A meeting of experts to recommend common nomenclature is scheduled to take place in 1978. MEBSS should not adopt any formal classification of mineral resource stocks

---

[†]This system of accounts is supplemented by consumer preference data, wage and price data, and the like.

[‡]All household activities—however "productive" they may be in actuality—are classified as consumption.

pending the outcome of the present coordinating effort.

As regards "renewable" natural resources, such as forests, fisheries, agricultural land, and so forth, the United Nations system has not yet adopted fully general classifications, although aquatic animals and plants, fishery commodities have been classified by the FAO and forest products have been classified by the FAO/ECE Timber Division. In this area, the most sensible procedure for MEBSS is to adopt the most appropriate classifications now in use in the United Nations system or in individual countries where relevant statistics are currently kept. The same thing applies to classifications of residuals and waste materials. Appropriate eight-digit code identifiers can presumably be developed that will be consistent with the ICGS framework.

As commented earlier, it is important for MEBSS to distinguish categories of material products not in terms of their relationship to conventional production/consumption concepts, but rather, in terms of their durability. In Figure 6.1 it is implied that all goods are neatly divided between "consumables" and "durables." This is vastly oversimplified and probably misleading, since durables range in useful lifetime from a century (or more) for dams, dikes, highways, and masonry structures to as little as three years for certain industrial tools and dies and the like. Bridges, power plants, houses, factories, process equipment and prime movers, aircraft, trucks and buses, cars, appliances, clothing, and so on range in between. It is not at all clear whether the various consumption and investment categories currently in use are optimal for purposes of MEBSS. No definitive answer to this question can be given at the present time.

There is one other area where a completely new classification scheme appears to be needed, for which there is no precedent as yet. This is the breakdown of industrial sectors to the "process" level. The practical definition of a materials/energy transformation process was discussed briefly in Chapter 5. The important criteria are as follows:

1. One (or more) principal inputs are transformed into one (or more) principal outputs.
2. Both principal inputs and principal outputs have definite market value, by virtue of being routinely exchanged in a market place at established prices.
3. There are no marketable intermediates produced and used within the process.

4. Both principal material/energy inputs and principal material/energy outputs can be packaged and transported over significant distances.
5. The process or activity is carried out in a distinct establishment on a commercially important scale and separate financial accounts are (or, at least, could be) kept for that particular establishment.

Evidently every commercial establishment incorporates one or more processes, depending on whether there are marketable[†] intermediates. The taxonomic problem is merely to compile a catalog of process descriptions from standard published sources until all economically significant materials/energy transformations can be classified. Insofar as the materials and chemical sectors are concerned, most of the relevant processes have already been defined (and described) in the chemical and metallurgical engineering literature, some of which is cited in Chapter 5.

In the manufacturing and service sectors, highly aggregated production processes will often be sufficient for MEBSS namely, "metal working," "assembly" or "packaging." In some cases, the appropriate process description will coincide reasonably well with an ISIC sector, as for example, oil and gas drilling, heavy construction, or transportation (by air, rail, pipeline, ship, or truck).

There is another type of process that must be considered, however. I am referring to processes that convert material goods and/or forms of energy into nonmaterial *services*. Notice that several of the criteria already listed for materials/energy transformations processes must be modified if the output is a service. First, if the service is generated by a durable good, there may *not* be an established market price for the service *per se*. The market-pricing mechanism may be applicable only to the service-producing durables (e.g., houses, automobiles, appliances, furnishings, clothing). By the same token, the service itself may not be transportable. Nor is it normal to keep financial accounts for service-generating equipment like cars or refrigerators in cases where the equipment itself is marketable and depreciable, but the service produced by the equipment is not routinely market priced or exchanged.

Service-generating processes are often (misleadingly) termed "end uses." These cut across the ISIC

[†]At least, in principle.

definitions of commercial and residential sectors, but are important from the standpoint of energy use:

- Space heating.
- Water heating.
- Air conditioning.
- Refrigeration.
- Cooking.
- Clothes drying.
- Illumination.

Each of these can obviously be broken down further to distinguish categorically different types of durable equipment that generate the services (e.g., oil/gas furnaces versus heat pumps). MEBSS should allow for this additional detail, even though the supporting data will be initially difficult to obtain.

6.3 TABULATIONS PROPOSED FOR MEBSS

The stock/flow data required to complete the scheme outlined previously can be displayed conveniently in four major tabulations. The first, Table 6.2, displays aggregated data in common physical units (such as metric tons, gross) on production, consumption, and stocks of resources at all stages of processing, beginning with natural resources and unprocessed commodities and proceeding through all intermediate forms to finished materials and classifiable products. Fuels and electricity (in appropriate units) are also included explicitly. There is one column for each resource or commodity, so that Table 6.2 has at least as many columns as there are listed commodities in the ICGS manual, which has been extended to include resources.

It is certainly possible, in principle, to subdivide Table 6.2 into several smaller tables. For instance, natural resources (not extracted), raw materials, and fuel (as extracted), and processed materials and processed forms of energy could be separated into separate tables for convenience. Similarly, natural resources could be subdivided into renewable and nonrenewable categories with separate tables for each. These subdivisions may ultimately be done for convenience, since the combined table will obviously be very large. However, any separations also introduce some degree of arbitrariness which seems unnecessary. Table 6.2 utilizes multiple resource stock categories (rows 2–5) to give MEBSS the flexibility to deal with alternative concepts. Revaluations of the various "reserve" accounts will often involve

balancing shifts from one category to another, due to new discoveries, technological changes, or price changes. Thus if ore prices rise the recoverable fraction of already proved reserves will increase. Exploration, on the other hand, may either increase or decrease the "possible" and "probable" categories; or the latter may increase at the expense of the former. Proven reserves can only be increased at the expense of "probable" reserves, as the latter are prospected in detail.

The major innovation in Table 6.2 is that use for conversion into other materials or commodities (row 15) is distinguished from use without further conversion (rows 16–18). Thus pelletized or sintered iron ore is used for conversion to pig iron, in a blast furnace. Similarly, pig iron is mostly converted to steel, though some goes to iron castings and forgings. Steel is generally produced by conversion of pig iron or scrap iron in an appropriate furnace (open hearth, basic oxygen, or electric arc). These facts are reflected in the table.

Domestic use as such is the complementary category. It can be divided between "intermediate" and "final" destinations (rows 16 and 18, respectively). Intermediate use of materials in manufacturing and services can be further broken down to separately identify "embodied" uses, as distinct from uses where the material (or fuel) is consumed immediately. (All of these distinctions were discussed in Chapter 3.) Examples of materials consumed at the secondary production stage and not embodied in a final product would include acids, solvents, fuels, lubricants, or agricultural chemicals; on the other hand, most metals, glass, cement, food products, wood, and fiber materials are "embodied" in goods —durable or otherwise. Evidently, row 17 is a subset of row 16.

It should be noted that direct accounting identities relate a number of rows in Table 6.2. These are listed as follows:

$$2 + 6 = 21$$
$$3 + 7 = 22 \quad \text{Opening stock} + \text{revalu-}$$
$$4 + 8 = 23 \quad \text{ation equals closing}$$
$$5 + 9 = 24 \quad \text{stock.}$$

$$10 + 11 + 12 - 13 = 14 \quad \text{Components of supply}$$
$$15 + 16 + 18 = 19 \quad \text{Components of demand (use)}$$
$$14 - 19 = 20 \quad \text{Definition of inventory change}$$

Table 6.2   Resource/Commodity Accounts

	Resources		Commodities	
			Intermediate	Finished
	Renewable	Exhaustible	materials	materials
Row account	resources	resources	and fuels	and fuels
1 Extended ICGS code[a]				
2 Opening reserve possible recoverable resource[b]				
3 Opening reserve probable recoverable resource[b]				
4 Opening reserve proved recoverable resource[b]				
5 Opening strategic stockpile[b]				
6 Added to "possible" reserve by revaluation[b]				
7 Added to "probable" reserve by revaluation[b]				
8 Added to "proved" reserve by revaluation[b]				
9 Added to strategic stockpile[b]				
10 Primary domestic extraction/production				
11 Secondary domestic recovery				
12 Imports				
13 Exports				
14 Domestic supply				
15 Domestic use for conversion				
16 Domestic use "as such," intermediate (all)				
17 Domestic use "as such," (embodied only)				
18 Domestic use "as such," final consumption				
19 Domestic consumption, total				
20 Inventory change on unaccounted for				
21 Closing "possible" reserve[b]				
22 Closing "probable" reserve[b]				
23 Closing "proved" reserve[b]				
24 Closing strategic stock[b]				

[a]Extended to include resources.
[b]Not applicable to commodities.

The purpose of Table 6.2 is to display gross accounting balances for each major natural resource or commodity that is produced on a large scale and exchanged in a market. In the case of renewable natural resources, such tables would provide a basic management tool, for example, for adjusting extrac- tion rates to maximize output without depleting the resource. In the case of nonrenewable resources, the current extraction rate would typically be measured against the rate of upward revaluation of stocks due to exploratory activity, price increases, or improve- ments of extraction technology. Limitations on current production and on the rate of expansion of current production of resources constitute an important input to resource/environmental management and economic forecasting models.

Commodity data is included in the same basic tabulation because it provides the necessary link between calls on basic resources and final demand for goods and services. Note that Table 6.2 would include production, consumption, import, and export figures for a number of dangerous materials such as nuclear fuels, DDT, chlorinated biphenyls, fluorocarbons, PVC, mercury, and cadmium. Again, this data would primarily be utilized in resource/environmental management and forecasting models.

The resource/commodity balance data in Table 6.2 will generally be derived annually from current specialized statistics on agriculture, fisheries, forestry, mining, manufacturing, and trade. Departments of parks, tourism, and urban affairs will also provide some data. While these standard statistical sources are not necessarily of uniformly high quality, the overall figures thus derived will, nevertheless, constitute a useful means of updating and projecting the more detailed materials/energy balance-by-process data tabulations, discussed next.

The second set of tabulations required for MEBSS is the process materials/energy input–output accounts for each conversion process, shown in Table 6.3. These accounts are closely related to the origin/destination accounts discussed next. This fact will also be of assistance in checking the accuracy of raw data that is provided, primarily by census of industrial enterprises. The input–output tables required for each process are closely related to the columns of the resource/residual matrix, defined in Fig. 4.8 and Equations (4.42) and (4.43) except that the latter are reduced to dimensionless coefficients expressed in units per unit of principal product. Also a universal nomenclature for all inputs and outputs is obviously required.

The extended ICGS code, including nontraded resource inputs (such as air and water) and waste residuals will be used to identify all inputs and outputs. The entries in Table 6.3 compiled by the Census of Manufactures would be obtained in terms of actual quantities purchased and produced by specific establishments. Much of the data on nonpurchased inputs and waste residuals produced will *not* be obtained from (or, perhaps, even known to) the individual industrial establishments. On the contrary, much of this supplementary data will have to

be filled in by independent engineering analysis using data from environmental monitoring surveys and similar sources. However, the materials balance principle applied separately for *each* chemical element offers a fairly powerful tool for uncovering inconsistencies and errors in data supplied by respondents.[†] As many readers are aware, industrial data of this kind is generally of low quality and respondents to census questionnaires are typically accountants, not engineers. Also, some firms may deliberately falsify their data (e.g., on waste emissions) to deceive regulatory agencies. Materials/energy balances cannot necessarily identify and correct all errors and discrepancies, but they can be very helpful to an auditor.

To complete the statistical system, it is convenient to treat both household and government or institutional consumption and depreciation of durables, including capital goods, as activities or processes. Also waste treatment, pollution abatement, and secondary materials recovery processes must be included. As regards the latter process categories, it almost goes without saying that the inputs are not in general purchased, since these activities are generally subsidized by government.

The purpose of Table 6.3 is to display moderately detailed relationships between inputs and outputs of materials and energy, including wastes, at the successive stages of production and transformation of natural resources into commodities and subsequently into final goods and services, and at last into refuse. Combined with data in the next set of tables (origin/destination accounts), it will ultimately provide a basis for projecting detailed resource and energy requirements and gross (preabatement) residuals generation, consequent to changes in the technological level of industry or the pattern of government regulation. It also permits (in principle) the application of mathematical programming algorithms to facilitate the optimal choice of industrial development strategy and pollution abatement strategy in terms of resource and conservation and environmental protection criteria.

Statistical data for Table 6.3 is not completely available at present, in any country, and would have to be gathered especially, primarily via the Census of Manufactures, Census of Households, or Census of

---

[†]The use of materials balances to uncover discrepancies and correct errors has been suggested by Peskin[9] and Avenhaus,[10] as well as Cummings-Saxton et al. (see Section 4.4).

Table 6.3   Materials/Energy Input–Output Accounts

Principal product name □   Process or activity name □
ICGS code   X   Process code   X

Column	Resource/commodity[a] input accounts			Product/waste[a] output accounts		
	Extended ICGS code	Name	Input quantity	Extended ICGS code	Name	Output quantity
	1	2	3	4	5	6
Materials						
Energy						

[a]Inputs listed in (extended) ICGS order, including nontraded resources and wastes.

Municipalities. This is unlikely to be carried out on an annual basis (in the forseeable future) and would, in any case, involve significant delays for data processing and checking. Thus Table 6.3 will normally be compiled several years later than Table 6.2 —being two or three (or more) years out of date when published—and it will be prepared at less frequent intervals. It will constitute a "benchmark" in the same sense as the input–output tables now prepared at intervals by several governments (and by the ECE).

The detailed balancing of inputs and outputs throughout the economy also requires a set of materials/energy origin/destination accounts. Specifically, there will be a destination account for each resource column shown in Table 6.1. All other commodities and waste residuals will have both origin and destination accounts. Both types are shown in Table 6.4. There are only three columns in the "origin" accounts for each commodity, but as many rows as there are identifiable originating domestic processes, plus a row for "imports and unaccounted for," the numbers in column 3 must add up to unity, by definition.

There are six columns in the "destination" accounts for each resource commodity or waste residuals, including consumption wastes and depreciation of durables. There will be exactly one row for each explicitly identified conversion process (process code), plus one for each explicitly identified sector destination (ISIC code) where no conversion is in-

volved. As noted earlier, an ISIC code is typically given by four digits corresponding to a group of commodities or services. There are ISIC code numbers for final consumption and investment categories, of course, and there is an extra "exports and unaccounted for" row to balance the accounts. This category also includes any year-to-year inventory changes.

Table 6.4 defines the links between economic activity—by production and consumption—and the production and disposition of waste residuals and pollutants. The statistics will be used, primarily, in forecasting the environmental consequences of economic growth and development, or in designing governmental programs to improve the state-of-the-environment or to minimize the adverse environmental impact of development projects. Destinations of treated (or untreated) waste flows are identified as to environmental media.

The basic source of these statistics will be some combination of engineering studies (e.g., of a "typical" plant exemplifying a given extraction or production process) and of survey or monitoring data compiled by responsible monitoring and regulatory agencies. These data are not currently available in adequate detail in any country,[†] and much attention will have to be given to the problem of designing

---

[†]A few countries are beginning to compile this kind of data, more or less haphazardly. The United States, Canada, Norway, The Netherlands, and Finland are probably the most advanced.

Table 6.4   Materials/Energy Origin/Destination Accounts

Material or commodity name

ICGS code

Origin accounts[a]			Destination accounts[b]					
Originating process code	Quantity produced by each process	Fraction produced by each process	Destination ICGS on ISIC code	Quantity	Fraction	Conversion process code	Quantity consumed by process	Fraction consumed by process
1	2	3	4	5	6	7	8	9
X	X	X	X	X	X	X	X	X
X	X	X				X	X	X
X	X	X				X	X	X
X	X	X				X	X	X
Imports and unaccounted	X	X	X	X	X	X	X	X
	Total production plus imports	1.00				X	X	X
						X	X	X
			X	X	X	X	X	X
						X	X	X
						X	X	X
			X	X	X	—	—	—
			X	X	X	—	—	—
			Inventory exports and unaccounted					
				Total consumption plus exports	1.00		See Row 15, Table 6.3	

[a]Listed in order of process importance.

[b]Listed as follows: conversions first, in order of ICGS importance: (by process, in order of process importance), followed by unconverted uses in order of ISIC importance.

[c]Conversion processes listed where applicable (i.e., where destination is specified by an eight digit ICGS Code).

appropriate methods of measurement and aggregation.

Table 6.4 could include such data as the rate of loss of mercury to the environment from the mercury-cell chlorine manufacturing process or the cadmium content of wastes from zinc mining and/or smelting. It would also, in principle, show mercury and cadmium residuals resulting from the incineration of paper impregnated with mercury-based fungicides or plastic products containing cadmium-based plasticizer. For DDT, PCB, and fluorocarbons, it would show waste flows resulting directly from "final" use of these materials.

The sum of all domestic production plus imports (column 2) is forced to be equal to the sum of all domestic consumption plus exports (column 5). The correspondence with Table 6.2 is maintained on the "supply" side; that is, column 2 of Table 6.3 must also be equal to the sum of rows 10, 11, and 12 in Table 6.2 (or, row 14 *minus* row 13). Any statistical discrepancies between known domestic supply and known domestic consumption are allocated to "inventory changes" in Table 6.2 and "inventory, exports and unaccounted for" on the destination side of Table 6.4.

The sum of all fractions in column 6 add up to unity, again by definition. The sum of all quantities in column 9 should add up to the "total domestic use for conversion" given by row 15 in Table 6.2, while the sum of all fractions given in column 9 should be equal to the ratio of the sum of column 8 to the sum of column 5, that is, the fraction converted. It is worth emphasizing that these interconnections between Table 6.2 and 6.3 will certainly be of significant value in verifying the more detailed accounts.

The relationship between Tables 6.3 and 6.4 can also be easily seen. The entries in Column 2 of Table 6.5 representing quantities *produced* by originating process correspond to products listed under "output quantity" (column 5) of Table 6.3 for each of the processes in question. Similarly, the entries under Column 8 of Table 6.4 representing quantities *consumed* by process correspond exactly to the "input quantity" heading (column 3) of Table 6.3. Thus most of the information in each table can be inferred from the other, except that Table 6.3 completes the materials/energy balance for each process by explicitly including all nonpurchased resource inputs and waste products, whereas Table 6.4 completes the balance for each commodity by explicitly including imports, exports, and uses of the commodity as such in unconverted form.

Dimensionless physical input–output coefficients by process and by commodity can be developed from Tables 6.3 and 6.4 by expressing all mass or energy per unit of principal mass or energy product, respectively. However, mixed dimensional coefficients—energy input per mass unit of output, or mass input per energy unit of output—require supplementary tables displaying the equivalences between energy content and mass (or volume[†]) for all fuels and other materials in which recoverable energy is embodied (such as scrap metal and waste paper). Typical equivalance relationships that must be specified include: density (mass/volume), energy content (Btu/kg or Btu/liter), and generating efficiency (kWh/Btu). Unfortunately, we must contend with the fact that terms in common use vary from year to year and from country to country.

The fourth and last table (Table 6.5, Accumulation Accounts) is concerned with production, consumption (depreciation), and stocks of durable goods, classified by ICGS category. The data in Table 6.5 will be recorded in "natural" units of account as well as in terms of gross weight (mass). Natural units would often be a simple integer tally, for example, for residential houses, passenger cars, or major appliances. However, other units may be appropriate, as in the case of highways or railroads (lane-miles or track-miles), commercial buildings (floor area), and so on. A supplementary table should also be developed, indicating typical or average unit weight and materials content of the various categories of durables.

The purpose of Table 6.5 is mainly to develop a record of stocks of durables, in-use, from which inferences can be drawn with regard to future waste generation, and materials potentially available for recycling, particularly metals. A gross accounting type of mass balance is likely to be adequate. Some —if not most—of the data, except on depreciation and scrappage, is already available from conventional sources. Data on depreciation and scrappage of durables will require some extension of current Census of Households and Census of Manufactures.

The table would also be useful in the event that some material already in widespread use was found to be potentially harmful (as, for instance, lead pipe

---

[†]Basic data for petroleum and natural gas is often given in volumetric terms.

Table 6.5   Durable Goods Accumulation Accounts

Row	Column / Name of category		1 Residential Houses	2 Commercial Buildings	3 Roads and Highways	4 Passenger Cars	5 Trucks and Buses
1	Extended ICGS code						
2	Units of account						
3	Opening	Units					
4	stock	Weight					
5	Domestic	Units					
6	output	Weight					
7	Imports	Units					
8		Weight					
9	Exports	Units					
10		Weight					
11	Added to	Units					
12	stock	Weight					
13	Scrappage	Units					
14		Weight					
15	Reval. of	Units					
16	stock	Weight					
17	Closing	Units					
18	stock	Weight					

or lead-based paint). Many household appliances or machines contain component elements that may cause environmental hazards when the equipment is eventually discarded, especially if it is incinerated. Relevant examples include fluorocarbon refrigerants, high-temperature coolants (based on PCB), mercury-vapor lamps, and mercury switches. The rapidly spreading use of synthetic fabrics and plastics for household furnishings and even structural components, combined with an enormous number of different plasticizers, stabilizers, fire retardants, coloring agents, water repellants, and so forth, raises strong possibilities that currently unsuspected risks will come to light from time to time. When this occurs, statistical data on the distribution and use of the dangerous substances will obviously be needed.

Before concluding, it may be useful to comment briefly, once again, on the treatment of uncertainty in the tables. There is no consistent way to characterize the intrinsic quality (as measured, for instance, by the variance) of much of the standard sources of statistical data available to governments. This statement is as true for the data used in the SNA as it is for the data that would be used in MEBSS. The only exceptions would be data obtained from scientific measurements under highly controlled conditions; government tax receipts and outlays; monetary flows through banks; data from a complete census (or administrative records of equivalent scope), or data from a large scientific sample survey. Incomes and expenditures for corporations or for households (or individuals) cannot be obtained directly with great confidence. Fortunately, accounting identities are available to ensure that various income, outlay, and accumulation accounts are internally self-consistent. This does not, however, ensure their accuracy in any absolute sense, of course.

In the case of materials and energy inputs, outputs, and accumulations, the same remarks apply. Data must be derived from a variety of sources, of various degrees of intrinsic quality. There is no way to characterize the certainty, or variance, of much of the data. But, again, the existence of accounting balances can be a great help in *reducing* the uncertainties.

It must be recognized that raw data for tables such as MEBSS can never be measured directly by a centralized statistical agency. It must always be obtained, initially, from questionnaires sent to firms or other agencies of government. Thus the role of the compiling agency is *not* to design and implement scientific methods of measuring process data, or the like. Nor could it even monitor the actual sampling or data collection. Its role is, rather, to design the system as a whole to minimize the individual importance of errors in measurements designed by others, or, for that matter, of human errors or even deceptions.

6.4 SCOPE, COST, AND IMPLEMENTATION STRATEGY

At first sight, the scheme of tables described in the previous section seems very ambitious. A closer examination of the problem provides little, if any, reassurance. The unvarnished fact is that the proposed MEBSS is a very large undertaking indeed. Key questions that deserve some explicit attention here include the following:

1. How many different resources, commodities, and residuals would be covered by it? How many separate processes would be distinguished?
2. For a given degree of coverage, how important are the omissions, that is, the items lumped together into "basket" categories?[†] What assurance can be given that the existence of such basket categories does not negate the remainder of the system by invalidating the materials/energy balances?
3. What are the criteria for inclusion of a commodity or process?
4. How good are the data?
5. Can the system be implemented by stages? Will an incomplete version have any value?

I will try to answer these questions in order. Beginning with the first, the question evidently assumes that there is some *a priori* definition of the scope of a statistical system. On closer reflection, however, it is clear that no such *a priori* definition is possible. The number of distinct categories must be determined by the needs of users of the system. To put the matter in perspective, the Canadian industry–commodity table[11] lists about 640 explicit commodities and service categories, of which only 209 or so are distinct basic nonagricultural materials, including 72 specific industrial chemicals.

On the other hand, in a recent input–output study of the Chemical Industry,[12] some 400 chemicals were identified. It is my best guess, at this time, that an adequate degree of detail would be obtained with

500 to 750 major chemicals and synthetic organic materials and no more than 1500 chemical processes. For the rest of the basic materials sectors, it seems likely that 200 categories of metals, identified as to type of alloy and configuration,[†] would be sufficient. Most of these would be explicitly limited to one or two alternative production processes, though a larger number of refining options might exist.

Food and nonfood agricultural commodities constitute a more serious difficulty, since there are a number of minor variations on cultivation processes and a wide spectrum of distinguishable but generically similar products such as feedstuffs, fruits, nuts, meat products, milk products, baked goods, and confectionaries. Classifications in this area tend to be more arbitrary, and there is always some aggregation involved. However, there are not very many fundamentally different processes for harvesting, preserving, and preparing food. In rough terms, the relevant processes are: washing; cutting; crushing or grinding; separating stems, leaves, seeds, shells, or bones; air drying; freeze-drying; freezing; baking; roasting; smoking; boiling; and blending or mixing, fermenting, and distilling. These processes are virtually identical regardless of the nature of the food, whether milk, fruit, grain, or fish. Similarly, widely different foods can be, and are, packaged in similar cans, bottles, or plastic containers.

This general situation holds, too, for much of the manufacturing, construction, and service sectors of the economy. A small number of cutting, forming, and fabricating processes are common to all structural materials. A very few assembly processes are similar in all manufacturing industries. A few processes for applying protective or decorative coatings are applicable to virtually all manufacturing industries from textiles to automobiles. A few modes of transportation suffice for all kinds of commodities. Beyond the basic materials, fuels, and energy transformation industries, it is not unrealistic in most cases to identify a sector with a single composite (multiestablishment) process.

From these observations, a rule and a conclusion can be drawn. For MEBSS, two materials or commodities should *not* be treated as an aggregated class, if they are produced or converted by clearly different processes. However, grouping similar prod-

---

[†]Usually labeled nec or "not elsewhere classified."

[†]By configuration, I mean ingots, billets, slabs, sheets (rolled), foil, castings, tube, wire, powder, and the like.

ucts together is relatively safe as long as the same process is used for all members of the group. It follows, then, that the scope and coverage of MEBSS must be governed by the number of clearly different significant processes in the economy.

The term "significant" deserves further explanation, since it is obvious that there are a few transformation processes of extreme economic importance, but a much larger number of less important processes and a vast profusion of minor ones. Presumably nobody would doubt that the processes of transforming crude oil into heating oil and gasoline or iron ore into pig iron and thence into steel are significant by any standard. But must the MEBSS system also necessarily distinguish different processes for casting bronze statues, manufacturing stained glass windows, or creating holes in Swiss cheese? If the answer to this admittedly rhetorical question is no—which it is—the obvious follow-on questions are: Where do we draw the line? or if that cannot be answered directly, How is the decision to be made?

The answer to the latter question depends on having a measure of the importance of a process in the economy. For materials and energy transformation processes, the simplest appropriate measure is value added by the process in relation to the value of the principal outputs. Thus if all the production processes in the United States' economy as a whole were identified and rank ordered, we would presumably wish to include (at least) the N top ones on the list, where N is probably a number greater than 1000 and less than 10,000. We would argue that the exact point of cutoff would be precisely where the marginal cost of adding an additional category to the system exceeds the marginal benefit of doing so.

This criterion, while theoretically valid, is useless in practice since the marginal costs and benefits are not calculable. Besides, marginal costs and benefits of adding categories to MEBSS are *not* simply functions of value added by process. It is easy to cite instances where value added *per se* is secondary, due to overwhelming environmental implications of the use of certain critical materials.

Thus on reflection, it would appear that, in addition to choosing the N highest ranking processes from an overall value-added standpoint, we should select processes in order to account for the flows of environmentally important materials or forms of energy, up to a specified level of precision. For instance, mercury and cadmium are economically "minor" metals, but environmentally they constitute major hazards. Thus it is quite likely that MEBSS would be designed to incorporate all industrial or consumption processes resulting in the release of environmentally significant quantities of these materials. The meaning of "significant" in this context is clearly in the province of environmentalists or health officials, not economists.

In addition to production processes already discussed, it is necessary to deal with consumption processes, including the various categories of durable goods and equipment. Residuals abatement and waste treatment processes must also be dealt with explicitly. In both of these cases, the value-added criterion is irrelevant (or, at least, unusable), and we must fall back on physical measures of significance. Again, the important thing is evidently to account accurately for the materials and energy flows, with particular emphasis on those that are environmentally most important. Thus more accurate data is required for toxic or carcinogenic waste flows than for merely bulky ones. It follows that more process differentiation may be needed to monitor the uses and disposition of chlorinated hydrocarbon than demolition wastes or yard trimmings.

I must add, here, that the decisions as to what materials and processes are included or excluded and what degree of accuracy of the data is "enough" are surely not going to be made by a group of MEBSS accountants in the UN Statistical Office or a national counterpart. These decisions will evolve over a period of time out of interactions between those who need and use data and those who collect and compile it.

In summary, I would estimate the "optimal" scope of MEBSS roughly as follows:

- 650 ± 150 Basic materials and fuels.
- 2000 ± 500 Materials/energy transformation processes associated with specific materials.
- 250 ± 50 Industrial sectors identified with aggregate commodities and composite processes.
- 100 ± 50 Consumption categories and types of service generating durable goods and equipment.
- 75 ± 25 Residuals and categories of waste.
- 200 ± 50 Abatement, waste treatment, and recycling processes.

It is not possible at this stage to estimate relative importance of processes that would still be aggregated into basket (nec) categories, or the resulting inaccuracies in materials or energy balances. I believe the omissions and resulting errors would be quantitatively minor—in the range of a few percent at most—but this conjecture cannot be verified until MEBSS is actually implemented.

As regards the quality of data, I have commented already that much of what is needed is not currently available. Unquestionably, this is partly because good data, for instance, on materials/energy balances for industrial processes is simply hard to get. On the other hand, the need of resource, energy, and environmental regulatory agencies clearly require such data, and in the past few years, in the United States alone, many millions of dollars have been spent gathering information on significant industrial processes. The logical starting point for MEBSS would be to utilize this mass of existing information —unreliable though it may be, in many cases—and subject it to the many tests of internal consistency that the materials/energy balance principle provides. These consistency checks will inevitably uncover numerous errors and discrepancies, which can be corrected later by means of appropriately well-specified surveys or engineering analysis. Thus while the first edition of MEBSS might be inaccurate in many details—as SNA data for many countries still is—the materials/energy balances at least guarantee a physically possible set of accounts. And as later and more accurate information becomes available, this same feature assumes that secondary (compensating) errors can be uncovered and corrected as well.

The final question to be discussed is whether MEBSS could be implemented by stages and whether a partial or incomplete version would still have significant value. My answer is, absolutely, yes. The best evidence of feasibility and usefulness is to present a meaningful example. This is done in the following section.

6.5 POSSIBLE SUBSETS OR SIMPLIFICATIONS OF MEBSS FOR INITIAL IMPLEMENTATION EXAMPLE: FLUOROCARBONS

In view of the vast scope of MEBSS in its "complete" (i.e., ultimate) form, it is highly desirable to segment MEBSS to permit gradual construction of the full system in smaller stages. Both horizontal and vertical slices are possible. A "horizontal" cut would begin with one reasonably well-defined stage of materials flow (say, reduction of metal ores to metal) and identify all the relevant flows into and out of that group of processes. This would be a relatively simple extension of production and trade statistics already available in many countries. The two added features would be specification of waste flows from and energy inputs to primary metallurgical processes (by process) and specification of subsequent utilization of each metal according to the involvement of further conversion processes (such as forging or electroplating). The system could not, of course, be "closed" in any sense and materials/energy balances could be struck only for the metallurgical activities *per se*.

In a similar manner, other sectors or groups of processes could be dealt with individually or in groups. Thus all imports and exports of raw materials, commodities, and finished goods could be compiled to obtain a balance for the "rest-of-world" sector. Balances for "consumption of nondurables" and "accumulation of durables" could be arrived at by a combination of direct survey data and indirect (synthetic) methods. The same is true for conversion industries and manufacturing industries. Here census-type data would have to be supplemented by engineering studies. The inputs and outputs to the *waste* sector (presumably left to the last) would then be completely specified. Of course, direct analysis by survey and engineering studies would be desirable to confirm the results of indirect analysis, or identify areas of unreliable data.

The weaknesses of this approach are clear: first, the piecemeal sector-by-sector assembly of statistics does not yield a very useful product until it has been completed. Moreover, some of the segments are still (perhaps) indigestibly massive statistical projects, even if undertaken separately.

The "vertical" approach offers more promise. Here we would identify a single element (or group of related materials) and follow *it*, individually, through the economy from raw material to final consumption, taking into all waste flows along the way. More and more elaborate networks can be built by combining a number of such chains. The larger the network covered, the more useful it will be —though still incomplete.

In my view, it would make sense to begin with a set of the most environmentally critical materials, such as toxic heavy metals (lead, cadmium, mercury, etc.). Most of these metals have a fairly restricted set of uses. Experience acquired in compiling the process waste, and consumption flow data on these metals would be helpful in dealing with the more widely used and more complex flow patterns of the major metals. Moreover, the data sets for many metals overlap significantly. For instance, cadmium is a coproduct of zinc refining. In compiling necessary data on cadmium, we would necessarily obtain quite a lot of relevant data on zinc, thus simplifying the task for zinc (to be undertaken later). Similarly, zinc and lead often occur together, along with copper, silver, arsenic, and other metals. The more data obtained for any of these, the easier it is to fill in the rest.

The logic of a vertical approach beginning with the more environmentally dangerous substances and then gradually extending to other materials can be also seen from the fact that some materials (or elements) are far more critical than others, both from conservationist and environmentalist standpoints. A disproportionate amount of environmental damage is caused by a small number of processes (e.g., combustion of fossil fuels) and by chemical compounds composed of a small number of specially potent elements.

In this context, it is noteworthy that the *major* elements of biological systems (hydrogen, carbon, oxygen) are relatively innocuous when combined with each other, but that four *minor* elements of living systems (nitrogen, sulfur, phosphorus, chlorine) combine with these three, singly or all together, to form a large variety of toxic or carcinogenic compounds from NO_x and SO_x to cyanides, mercaptans, nitrosamines, nerve gases, such as CSN or phosgene, peracyl acetic nitrate (PAN), bacteriocides, algicides, fungicides, herbicides, insecticides, food preservatives, vinyl chloride, and polychlorinated biphenyls (PCBs). Clearly, monitoring these four elements and their compounds should take precedence over monitoring compounds that contain only hydrogen, carbon, and oxygen.

Similarly, a group of minor metals, including lead, cadmium, mercury, arsenic, and chromium, are far more dangerous to the environment than the more widely used elements such as compounds or alloys of aluminum, calcium, iron, silicon, and sodium. Again, even though the latter are far more important, economically, priority in monitoring should go to the former.

To conclude this discussion (and the book), it is instructive to provide a real-life example. The network I have selected to illustrate the structure of

MEBSS is centered around two key fluorocarbon compounds (CCl_3F and CCl_2F_2), more familiarly known as F-11 and F-12, respectively. These compounds, which are used as aerosol propellants, refrigerants, and "foam-blowing" agents, are of interest to environmentalists because it has been learned quite recently[13] that they remain for very long times (> 30 years) in the atmosphere. This is a long enough time to diffuse gradually into the stratosphere, above 25 km in altitude. In that rarified region, the flux of ultraviolet radiation from the sun is very intense. It tends to break down ordinary oxygen molecules (O_2) into atomic oxygen. Some of the latter recombines with ordinary oxygen to form ozone: $O_2 + O \rightarrow O_3$. Ozone, in turn, happens to be a very good absorber of ultraviolet light; it acts as a "shield" to prevent excessive levels of biologically dangerous ultraviolet radiation on the Earth's surface.

As it happens, when F-11 and F-12 reach the stratosphere they, too, are dissociated by ultraviolet light yielding (among other things) chlorine atoms. Unfortunately, the chlorine atoms react with ozone in the sequence:

$$Cl + O_3 \rightarrow Cl\!-\!O + O_2$$

$$Cl\!-\!O + O \rightarrow Cl + O_2$$

whence, combining the two stages

$$Cl + O_3 + O \rightarrow Cl + O_2 + O_2$$

In effect the chlorine is a catalyst for the reduction of ozone and monatomic oxygen back to ordinary molecular oxygen. One chlorine atom can destroy thousands of ozone molecules without being removed itself. It is for this reason that a ban on the use of F-11 and F-12 as an aerosol propellant is under serious consideration by the U. S. government.[14–16]

In view of the foregoing remarks, it is obvious that an important segment of the chemical industry is threatened. However, it is not at all clear how far the ban should extend or how quickly it should be imposed. It might be supposed that a ban on the use of certain fluorocarbon gases for foam insulation, aerosol deodorants, and hairsprays would impose no very serious loss to society. Be that as it may,† it is

---

†For purposes of this discussion, I neither endorse nor oppose this view.

also clear that fluorocarbons are by far the best available refrigerants and to deprive society of refrigeration (and air-conditioning) would indeed be a major step, not to be taken lightly.

Having said this, it follows that a detailed analysis of all the direct and indirect costs of banning fluorocarbons is a prerequisite to any rational decision. Even before an economic study can be undertaken, a detailed specification of the affected sectors must be made. I will now show that such a specification is tantamount to a vertical "slice" of MEBSS, in precisely the sense discussed previously.

Clearly, the manufacturers of F-11 and F-12 cannot be considered alone, since each of these chemicals is manufactured from intermediates. In several cases, very large fractions of the intermediate chemical is destined for fluorocarbons. This is notably true for carbon tetrachloride (100%), chloroform (93%), carbon disulfide (32%), and hydrofluoric acid (44%). These chemicals, in turn, account for significant fractions of the output of various other precursor chemicals and raw materials such as fluorospar, chlorine and sulfuric acid. The complete system of precursors and chemical processes involved is shown in Figure 6.2. A detailed discussion of the derivation of the figures is given in Appendix 6 A, together with a set of tables similar to Tables 6.3 and 6.4.

The correspondence between the tables and the flow chart can be seen at a glance. Materials are denoted by circles; processes by squares. The number in the circles represent total domestic production in 1973, in millions of pounds.[†] The arrows directed into a circle represent alternate sources, and the fractional amount derived from each source is indicated. (A source may be a specific process or, possibly, a broader category such as "imports.") Similarly, the arrows directed away from each circle represent the possible destinations. Again, the fractional allocation among different destinations is indicated on the chart. Tables 6-A.1 (1 to 11) are specific illustrations of the generalized Table 6.3. There is one such table for each *process* (box) in Figure 6.2. Tables 6-A.2 (1 to 6) are similarly, specific illustrations of generalized Table 6.4. There is one for each *material* (circle) in Figure 6.2.

The small subsystem of MEBSS illustrated in Figure 6.2 and the tables in Appendix 6-A is very helpful for any comprehensive environmental–economic–technological analysis of the impact (for instance) of banning certain fluorocarbon products. The materials flow network illustrated in the chart has its economic counterpart, as shown in Figure 6.3, where the 1973 price of each chemical is displayed in its circle, the dollar value of each material flow is noted and the total value added by each process is displayed in the square boxes. In words: the value added by process (or activity) is the difference between the cost of all purchased inputs and the revenues obtained from products sold. Here, too, the accounting balances must always be maintained even if the available data is inadequate. The correspondence between the materials balance scheme and the conventional input–output system is clearly illustrated.

REFERENCES

1. W. Leontief, "Theoretical Assumptions and Nonobserved Facts," *Econ. Rev.* (March 1971).
2. R. U. Ayres, "Preliminary Description of Materials/Energy Balance Statistical Subsystem (MEBSS)," IRT-358-R, Draft Working Paper for United Nations Statistical Office, New York, January 1976.
3. "A System of National Accounts," ST/STAT/Ser.F/2Rev. 3, United Nations, New York, November 1969.
4. Publication No. E.68.XVII.8, United Nations, New York.
5. "Statistical Classifications Draft International Standard Classification of All Goods and Services (ICGS)," Report of the Secretary-General, Part I, June 17, 1975; Part II, United Nations, June 12, 1974.
6. "Statistical Classifications Draft Standard International Trade Classification (SITC)," Rev. 2, Note by the Secretary-General, United Nations, New York, May 28, 1974.
7. "Explanatory Notes to the Brussels Nomenclature," United Nations, Customs Co-opeartion Council, Brussels, 1966.
8. J. Zwartendyk, "Interpretation of Data on Mineral Resources, Production and Consumption—Problems in Terminology," Background Working Paper for the United Nations Center for Natural Resources, Energy and Transport, New York, January 23, 1976.
9. H. Peskin, "National Accounting and the Environment," Artikler Fra Statistisk Sentralbyra Nr. 50, Oslo, 1972.
10. R. Avenhaus, "Materials Accountability, Theory, Verification and Applications," International Institute of Applied Systems Analysis, Laxenbourg, Austria, mimeo, July 1976.
11. "Users Guide to Statistics Canada Structural Economic Models" (Rev), Statistics Canada, Ottawa, February 1976.
12. J. Cummings-Saxton et al., "The Economic Impact of Water Pollution Control Costs Upon the Chemical Processing Industries," IRT-413-R, prepared for the National Commission on Water Quality, February 1976.

[†]As noted earlier, these numbers should obviously be presented into metric units for an international system.

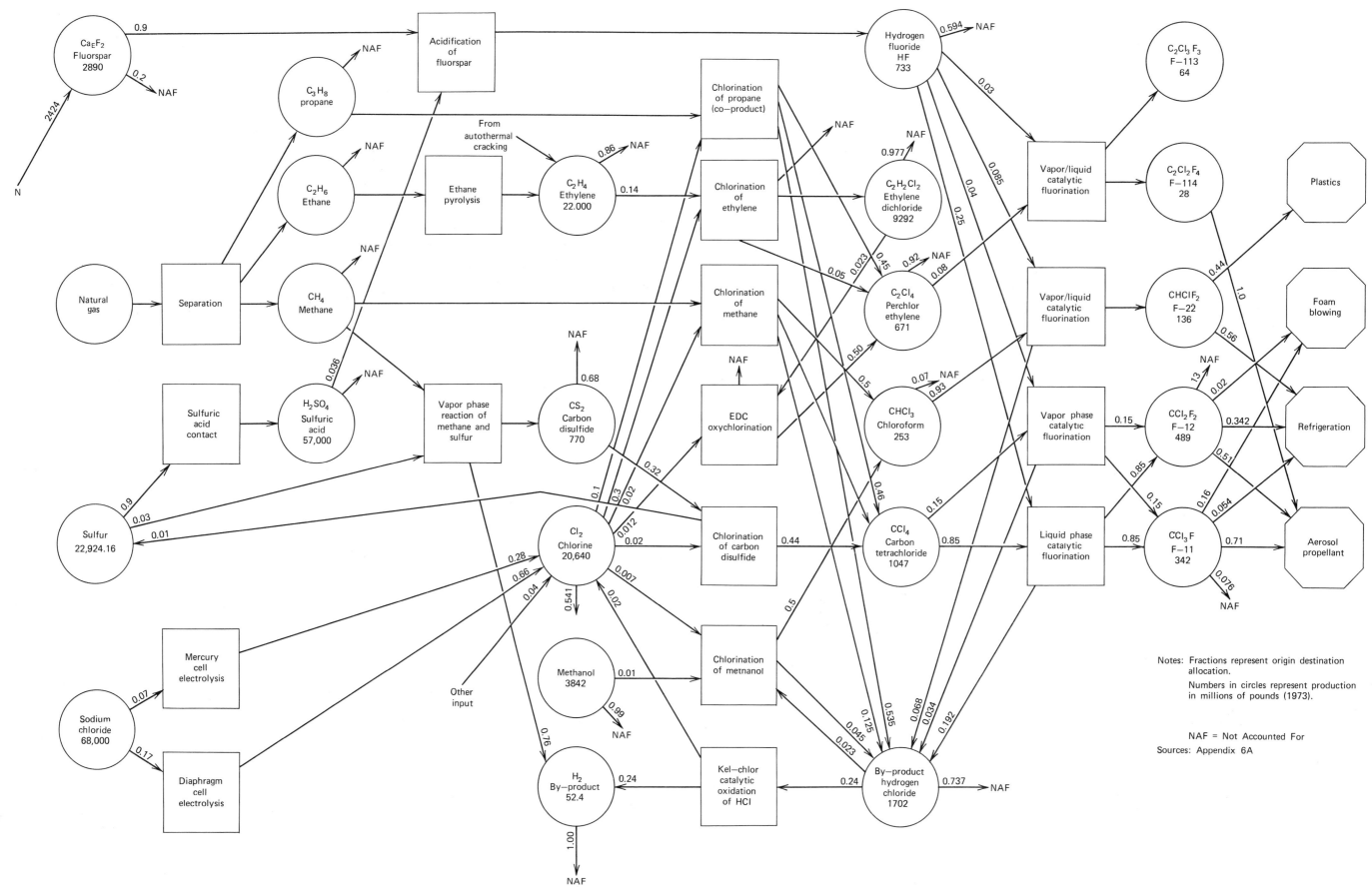

Figure 6.2. Chemical production and flows in fluorocarbon industry (U.S., 1973).

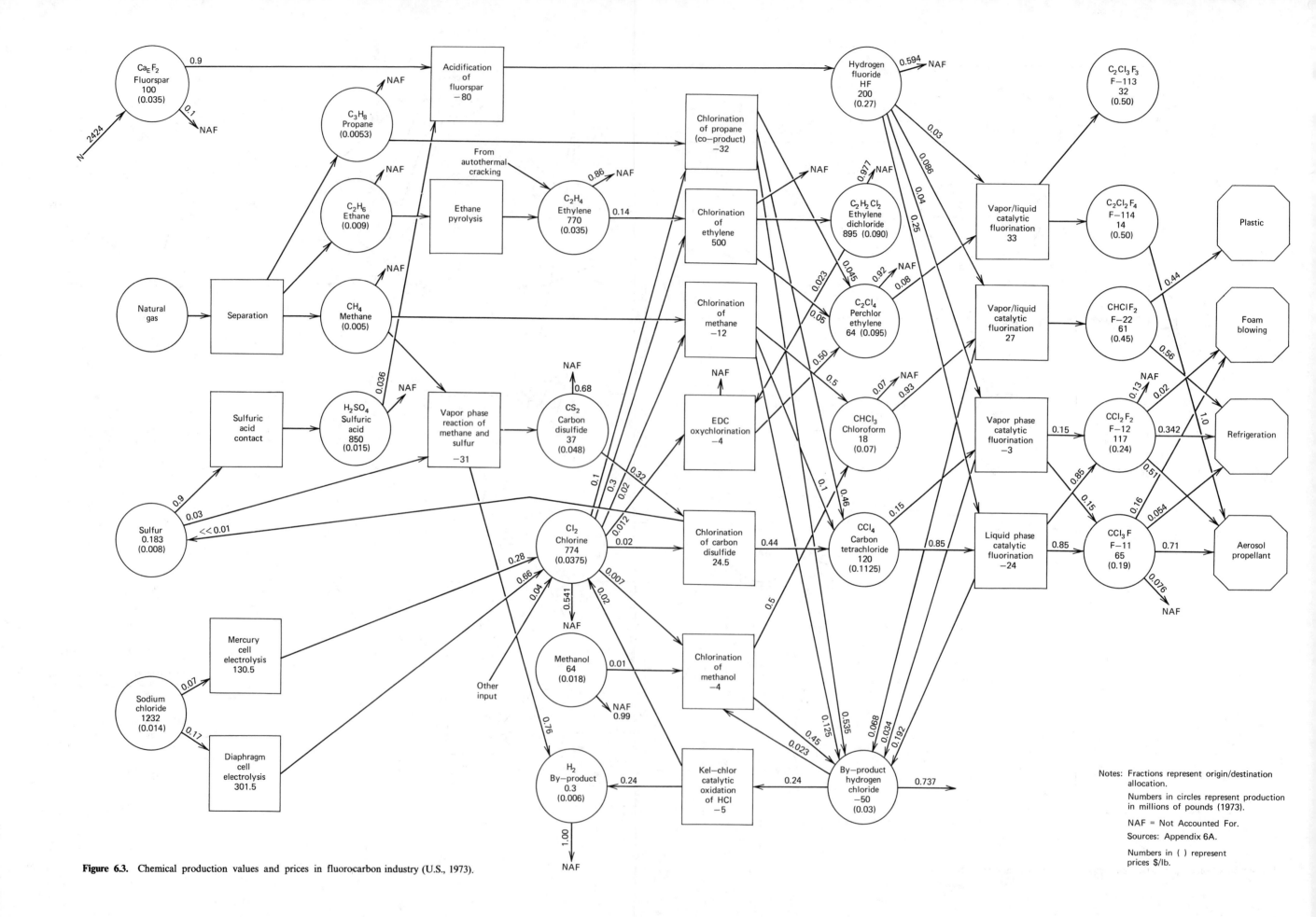

Figure 6.3. Chemical production values and prices in fluorocarbon industry (U.S., 1973).

13. M. J. Molina and F. S. Rowland, "A Stratospheric Sink for Chlorofluoromethanes: Chlorine Atom-Catalyzed Destruction of Ozone," *Nature*, **249** (June 1974).

14. Interdepartmental Committee for Atmospheric Sciences (ICAS), "The Possible Impact of Fluorocarbons and Halocarbons on Ozone," U.S. Federal Council for Science and Technology, Washington, DC, May 1975.

15. Interagency Task Force on Inadvertent Modification of the Stratosphere (IMOS), "Fluorocarbons and the Environment," U.S. Council on Environmental Quality and Federal Council for Science and Technology, Washington, DC, June 1975.

16. U.S. National Academy of Sciences, National Research Council, "Interim Report of the Panel on Atmospheric Chemistry, Climatic Impact Committee," July 1975.

APPENDIX 6-A[†]

DERIVATION OF NUMBERS IN FLOW CHART, FIGURE 6.2

The following chemicals' outputs are listed with their references:

- Sulfur, sodium chloride, and fluorospar from Ref. A1
- Ethylene dichloride, perchlorethylene, chloroform, and carbon tetrachloride from Ref. A2
- Sulfuric acid, carbon disulfide, chlorine, and hydrogen fluoride from Ref. A5
- Ethylene, F-114, F-11, F-12, and F-22 from Ref. A6.

All of the chemical equations used in this appendix were taken from *Faith, Keyes, and Clark's Industrial Chemicals*[A5] or a similar authoritative source. Balanced chemical equations are used to find quantities and distributions of those chemicals that were not defined elsewhere. The quantities of chemicals used in a chemical reaction must form a ratio equal to the ratios of the molecular weights multiplied by the chemical coefficient. Ratios of this nature were used to solve for unknown quantities.

The following chemicals, of which the origins or distribution are known, are listed with the corresponding reference: chlorine origins,[A4] carbon tetrachloride origins and distribution,[A3] ethylene distribution,[A5] carbon disulfide distribution,[A3] ethylene

[†]This appendix was prepared by Eric Weinstein.

dichloride distribution,[A6] perchlorethylene origins,[A6] and distribution.[A5] All chemical quantities, unless otherwise specified, are in millions of pounds.

The Uses of HF

In order to determine the percentage of HF output used in the production of each fluorocarbon, balanced chemical equations were used to find the quantity of HF necessary to produce the given amount of fluorocarbon, as follows:

F-114. The quantity of HF used in F-114 production is estimated from the reaction for F-114 production $4HF + C_2Cl_4 \rightarrow C_2Cl_2F_4 + 2HCl + H_2$.

The weights of inputs and outputs are:
mol wt $HF \times 4 = 80$ HF input = U (unknown)
mol wt F-114 = 171   F-114 = 26 million lb
The ratio is $80/171 = U/26$, whence

$$U = 12.2 \text{ million lb}$$

The C_2Cl_4 needed can be determined by the following:
mol wt $C_2Cl_4 = 166$ C_2Cl_4 input = U (unknown)
mol wt F-114 = 171   F-114 = 26 million lb.

The ratio is $166/171 = U/26$, whence

$$U = 25.2 \text{ million lb}$$

F-113. The quantity of HF used in F-113 production is estimated by making the arbitrary assumption that all of the C_2Cl_4 not used for catalytic fluorination to produce F-114 was used to make F-113.

Total C_2Cl_4 used for catalytic fluorination = 0.08 $\times 371 = 53.7$ million lb
Total C_2Cl_4 used for F-114 production = 25.2 million lb
The difference is 28.5 million lb for F-113.

The reaction for F-113 production is

$$3HF + C_2Cl_4 \rightarrow C_2Cl_3F_3 + HCl + H_2$$

The weights of inputs and outputs are:
mol wt $HF \times 3 = 60$ HF input = U (unknown)
mol wt $C_2Cl_4 = 166$ C_2Cl_4 input = 28.5
                                                     million lb
The ratio is $60/166 = U/28.5$, whence

$$U = 10.3 \text{ million lb}$$

F-22.
The reaction for F-22 production is

$$2HF + CHCl_3 \rightarrow CHClF_2 + 2HCL$$

The weights of HF and F-22 are:
  mol wt HF $\times$ 2 = 40 HF input = U (unknown)
  mol wt F-22 = 86.4   F-22 output = 136
$$\text{million lb}$$
The ratio of weights is 40/U = 86.4/136, so
$$U = 62.9 \text{ million lb}$$
U as a fraction of total HF = 62.9/733 = 0.086.

F-12.
The reaction for F-12 production is
$$2HF + CCl_4 \rightarrow CCl_2F_2 + 2HCl$$

The weights of HF and F-12 are:
  mol wt HF $\times$ 2 = 40 HF input = U (unknown)
  mol wt F-12 = 121   F-12 output = 489
$$\text{million lb}$$
The ratio of weights is 40/U = 121/489, so
$$U = 161.7 \text{ million lb}$$

F-11.
The reaction for F-11 production is
$$HF + CCl_4 \rightarrow CCl_3F + HCl$$

The weights of HF and F-11 are:
mol wt of HF = 20   HF input = U (unknown)
mol wt of F-11 = 137.5 F-11 output = 342
$$\text{million lb}$$
The ratio of weights is 20/U = 137.5/342, so
$$U = 49.7 \text{ million lb}$$
HF used for F-11 and F-12 sum = 211.4 million lb, so 211.4/733 = 0.29 of HF is used for vapor and liquid phase processes to produce F-11, F-12.

The Uses of Cl$_2$

The following equations were used to determine Cl_2 necessary for production of the given quantity of fluorocarbon precursor.

The reaction for CCl$_4$ production is
$$2Cl_2 + CS_2 \rightarrow CCl_4 + 2S$$
The weights of Cl$_2$ and CCl$_4$ are:
  mol wt of $Cl_2 \times 2 = 142$ Cl_2 input = U
$$\text{(unknown)}$$
  mol wt of $CCl_4 = 154$ CCl_4 output quantity =
$$460.7 \text{ million lb}$$
The ratio of weights is 142/U = 154/460.7, so
$$U = 424.8 \text{ million lb}; 0.02 \text{ of total chlorine}$$
$$\text{is used for } CS_2 \text{ process in making } CCl_4$$

The reaction for chloroform production is
$$3Cl_2 + CH_4 \rightarrow CHCl_3 + 3HCl$$

The weights of Cl$_2$ and chloroform are:
  mol wt of $Cl_2 \times 3 = 213$ Cl_2 input = U
  mol wt of $CHCl_3 = 119.5$ $CHCl_3$ output = 253
$$\text{million lb}$$
The ratio of weights is 213/U = 119.5/126.5, so
$$U = 225.5 \text{ million lb of } Cl_2$$

The reaction for CCl$_4$ production is
$$4Cl_2 + CH_4 \rightarrow CCl_4 + 4HCl$$

The weights of Cl$_2$ and CCl$_4$ are:
mol wt of $Cl_2 \times 4 = 284$
  wt of Cl_2 = U
  mol wt of $CCl_4 = 154$
  wt of CCl_4 product = 104.7 million lb
  The ratio of weights is 284/U = 154/104.7, so
$$U = 193 \text{ million lb}$$
Adding the above together, we obtain 418.5 million lb of Cl_2 needed for chlorination of methane; this is 0.02 of total chlorine output.

Cl_2 necessary for coproduct process is based on a ratio derived from Ref. A3 as follows:
$$4.18.2 \text{ million lb } Cl_2/U = 100/480, \text{ so}$$
$$U = 2007 \text{ million lb, or } .10 \text{ of total}$$
$$\text{chlorine output.}$$

The reaction for CHCl$_3$ production from methanol is
$$CH_3OH + HCl + 2Cl_2 \rightarrow CHCl_3 + 2HCl + H_2O$$

The weights of Cl$_2$ and CHCl$_3$ are:
  mol wt of $Cl_2 \times 2 = 142$ wt of Cl_2 = U
  mol wt of $CHCl_3 = 119.5$ wt of $CHCl_3 = 126.5$
$$\text{million lb}$$
The ratio of the weights is 142/U = 119.5/126.5, so
$$U = 150.3 \text{ million lb of } Cl_2 \text{ or } .007 \text{ of total}$$
$$\text{chlorine produced}$$

Sources of By-product HCl

HCl by-product comes from fluorocarbon precursor production. The previous two sections account for all reactions where HCl is given off. The following shows the HCl from each reaction:

From F-22 production the ratio of the weights is
$$86.5/73(2 \times \text{mol wt HCl}) = 136 \text{ million lb}/U, \text{ so}$$
$$U = 115 \text{ million lb of HCl or of by-product}$$

From F-12 production the ratio of the weights is
$$121/73(2 \times \text{mol wt HCl}) = 489 \text{ million lb}/U, \text{ so}$$
$$U = 295 \text{ million lb of HCl.}$$

From F-11 production the ratio of weights is
$$137.5/36.5 = 342 \text{ million lb/U, so}$$
$$U = 90.8 \text{ million lb of HCl.}$$

Summing the above two contributions yields 385.8 million lb of by-product HCl.

From $CHCl_3$ by CH_4 chlorination the ratio of weights is

$$119.5/109.5(\text{mol wt HCl} \times 3) = 126.5 \text{ million lb/U,}$$
so
$$U = 115.9 \text{ million lb HCl.}$$

From CCl_4 by CH_4 chlorination the ratio of weights is

$$154/142(\text{mol wt HCl} \times 4) = 104.7 \text{ million lb/U, so}$$
$$U = 96.5 \text{ million lb HCl.}$$

Summing the above two sources yields 212.4 million lb of by-product HCl.

From co-product process (ratio from Table C2[A3])

$$100/189.9 = 480 \text{ million lb/U, so}$$
$$U = 911.5 \text{ million lb HCl or } 0.536 \text{ of by-product HCl output}$$

From $CHCl_3$ by CH_3OH chlorination the ratio of the weights is

$$119.5/73(\text{mol wt HCl} \times 2) = 126.5 \text{ million lb/U, so}$$
$$U = 77.3 \text{ million lb, or } 0.045 \text{ of by-product HCl.}$$

Summing all the above, we obtain a total output of by-product HCl equal to 1702 million lb/yr.

Sulfur Content in CS_2

The reaction for CS_2 production is
$$2S + CH_4 = CS_2 + 2H_2$$
$$64(\text{mol wt S} \times 2)/76 (\text{mol wt } S_2) = U/770 \text{ million lb}$$
$$(\text{wt of } CS_2 \text{ output}), \text{ so}$$
$$U = 648 \text{ million lb or } 0.03 \text{ of total sulfur produced}$$

NaCl Output Uses

The NaCl used for chlorine production was derived using the structural relationship given in Reference A4. It was then converted into a fraction of total NaCl.

HCl By-product Uses

Of the by-product HCl that can be accounted for, 0.24 was assumed delivered to Cl_2 production by Kel-Chlor process in order to make it compatible with the Kel-Chlor listing in Source A4, and 0.023 was assumed used for methanol chlorination to account for the HCl used in that process.

H_2 By-product Sources

These values are taken from the balanced CS_2 equation and the Kel-Chlor process.

From CS_2 synthesis:

$$76(\text{mol wt } CS_2)/4(\text{mol wt } H_2 \times 2) = 770 \text{ million lb}$$
$$(\text{wt of } CS_2/U \text{ (wt of } H_2)), \text{ so}$$
$$U = 40.5 \text{ million lb of by-product } H_2.$$

From Kel-Chlor process:

$$71 (\text{mol wt HCl} \times 2)/2 (\text{mol wt } H_2) = 421 \text{ million lb}$$
$$\text{of HCl/U, so}$$
$$U = 11.9 \text{ million lb of } H_2.$$

Summing the above, 52.4 million lb of by-product H_2 were produced

Allocation of Inputs and Outputs to Processes Yielding F-11, F-12

This division was made according to the capacity available for each process as of 1975 as listed in Reference A3. This liquid phase process accounted for 0.85 of apparent capacity in that year.

REFERENCES

A1. *Minerals Yearbook*, U.S. Bureau of Mines, Washington, DC, 1973.

A2. *Synthetic Organic Chemicals: U.S. Production and Sales, 1973*, U.S. Tariff Comm., Washington, DC.

A3. J. Cummings-Saxton, M. E. Weber, R. U. Ayres, J. P. Merrill, and H. W. Pifer III. "Economic Impact of Potential Regulation of Chlorofluorocarbon Propelled Aerosols," IRT-462-R, April 1977.

A4. R. U. Ayres, J. Cummings-Saxton, and M. O. Stern, "The Materials–Process–Product Model," IRT-305-FR for NSF, July 1974.

A5. F. A. Lowenheim and M. K. Moran, *Faith, Keyes and Clark's Industrial Chemicals*, 3rd ed., Wiley, New York, 1975.

A6. Arthur D. Little, Inc., "Preliminary Economic Impact Assessment of Possible Regulatory Action to Control Atmospheric Emissions of Selected Halocarbons," for EPA, ADL-76072-80, September 1975.

Table 6-A.1(1) Materials/Energy Input–Output Accounts

Principal product name: CS_2 ICGS code: X
Process or activity name: [ ] Process code: X

	Resource/commodity input accounts[a]			Product/waste output accounts[a]		
Column	Extended ICGS code	Name	Input quantity	Extended ICGS code	Name	Output quantity[b]
	1	2	3	4	5	6
Energy Materials		CH_4	23.50		CS_2	100
		S	92.5		H_2S	8.8
					CH_4	2.4
					H_2	4.8

}Waste[c]

[a]Inputs listed in (extended) ICσS order, including nontraded resources and wastes.
[b]Per 100 lb of principal product.
[c]Figures estimated from efficiency of given industrial process.

Source: Reference A5.

Table 6-A.1(2) Materials/Energy Input–Output Accounts

Principal product name: CCl_4 ICGS code: X
Process or activity name: Co-product process Process code: X

	Resource/commodity input accounts[a]			Product/waste output accounts[a]		
Column	Extended ICGS code	Name	Input quantity	Extended ICGS code	Name	Output quantity[b]
	1	2	3	4	5	6
Materials Energy		C_3H_8	29.7		CCl_4	100.
		Cl_2	368.0		C_2Cl_4	107.8
					HCl	189.9

[a]Inputs listed in (extended) ICGS order, including nontraded resources and wastes.
[b]Per 100 lb of principal product.

Source: Reference A3.

Table 6-A.1(3) Materials/Energy Input–Output Accounts

Principal product name: CCl_4 ICGS code: X

Process or activity name: [blank] Process code: X

Column	Resource/commodity input accounts[a]			Product/waste output accounts[a]		
	Extended ICGS code	Name	Input quantity	Extended ICGS code	Name	Output quantity[b]
	1	2	3	4	5	6
Materials		CS_2	55		CCl_4	100.
		Cl_2	115		CS_2	5.7 }Waste[c]
					S	64.3
Energy						

[a] Inputs listed in (extended) ICGS order, including nontraded resources and wastes.
[b] Per 100 lb of principal product.
[c] Figures estimated from efficiency of given industrial process.

Source: Reference A5.

Table 6-A.1(4) Materials/Energy Input–Output Accounts

Principal product name: FC 11/12 ICGS code: X

Process or activity name: Liquid phase Process code: X

Column	Resource/commodity input accounts[a]			Product/waste output accounts[a]		
	Extended ICGS code	Name	Input quantity	Extended ICGS code	Name	Output quantity[b]
	1	2	3	4	5	6
Materials		CCl_4	125.85		VC 11	30
		HF	28.37		FC 12	70
					HCl	5.46
					HCl	1.03
					HF	0.11 }Waste
					FC 11/12	2.62
Energy						

[a] Inputs listed in (extended) ICGS order, including nontraded resources and wastes.
[b] Per 100 lb of principal product.

Source: Reference A2.

Table 6-A.1(5)   Materials/Energy Input–Output Accounts

Principal product name	FC 11/12	Process or activity name	Vapor phase
ICGS code	X	Process code	X

| | Resource/commodity input accounts[a] | | | Product/waste output accounts[a] | | |

Column	Extended ICGS code	Name	Input quantity	Extended ICGS code	Name	Output quantity[b]
	1	2	3	4	5	6
Materials		CCl_4	128.00		FC 11	30
		HF	29.18		FC 12	70
					HCl	50.39
					FC 13	1.5
					HF	0.07
					HCl	2.65
					FC 11/12	2.57
Energy						

(brace) Waste (for FC 13, HF, HCl, FC 11/12)

[a]Inputs listed in (extended) ICGS order, including nontraded resources and wastes.
[b]Per 100 lb of principal product.

Source: Reference A2.

Table 6-A.7(6)   Materials/Energy Input–Output Accounts

Principal product name	Cl_2	Process or activity name	Kel-Chlor
ICGS code	X	Process code	X

| | Resource/commodity input accounts[a] | | | Product/waste output accounts[a] | | |

Column	Extended ICGS code	Name	Input quantity	Extended ICGS code	Name	Output quantity[b]
	1	2	3	4	5	6
Materials		HCl	103		Cl_2	100
		H_2O	13.7		H_2	2.8
		N_0OH	0.05		HCl (gas)	0.004
					H_2O	1.0
Energy		kWh	80.6			

(brace) Waste (for H_2, HCl (gas), H_2O)

[a]Inputs listed in (extended) ICGS order, including nontraded resources and wastes.
[b]Per 100 lb of principal product.

Source: Reference A4.

Table 6-A.1(7)   Materials/Energy Input–Output Accounts

Principal product name: CHCl₃ → $CHCl_3$
ICGS code: X

Process or activity name: Methane chlorination
Process code: X

Resource/commodity input accounts[a]   Product/waste output accounts[a]

Column	Extended ICGS code	Name	Input quantity	Extended ICGS code	Name	Output quantity[b]	
	1	2	3	4	5	6	
Energy Materials		Cl_2	178		$CHCl_3$	100	} Waste[c]
		CH_4	13.3		HCl	83.95	
					Cl_2	7.35	

[a]Inputs listed in (extended) ICGS order, including nontraded resources and wastes.
[b]Per 100 lb of principal product.
[c]Figures estimated from efficiency of given industrial process.
Source: Reference A5.

Table 6-A.1(8)   Materials/Energy Input–Output Accounts

Principal product name: H_2SO_4
ICGS code: X

Process or activity name: Contact
Process code: X

Resource/commodity input accounts[a]   Product/waste output accounts[a]

Column	Extended ICGS code	Name	Input quantity	Extended ICGS code	Name	Output quantity[b]
	1	2	3	4	5	6
Energy Materials		S	34.4		H_2SO_4	100
		H_2O	1670		H_2O (waste)[c]	1604.4
		H_2O (steam)	10			
		kWh	0.25			

[a]Inputs listed in (extended) ICGS order, including nontraded resources and wastes.
[b]Per 100 lb of principal product.
[c]Figure estimated from efficiency of given industrial process.
Source: Reference A5.

Table 6-A.1(9) Materials/Energy Input–Output Accounts

Principal Product name	Cl_2	Process or activity name	*Mercury cell*
ICGS code	X	Process code	X

Resource/commodity input accounts[a] — Product/waste output accounts[a]

Column	Extended ICGS code	Name	Input quantity	Extended ICGS code	Name	Output quantity[b]
	1	2	3	4	5	6
Energy Materials		H_2O	449		Cl_2	100
		N_aCl	167.3		H_2	2.8
		HCl	0.9		N_aOH	96.0
		N_aOH	0.3		H_2SO_4 (dilute)	6.1
		H_2SO_4	6.1		CO_2	0.2
					H_2O	287
		kWh	165			
		H_2O (steam)	102			

(Waste: H_2, N_aOH, H_2SO_4 (dilute), CO_2)

[a]Inputs listed in (extended) ICGS order, including nontraded resources and wastes.
[b]Per 100 lb of principal product.
Source: Reference A4.

Table 6-A.1(10) Materials/Energy Input/Output Accounts

Principal product name	Cl_2	Process or activity name	Diaphragm cell
ICGS code	X	Process code	X

Resource/commodity[a] input accounts — Product/waste[a] output accounts

Column	Extended ICGS code	Name	Input quantity	Extended ICGS code	Name	Output quantity[b]
	1	2	3	4	5	6
Energy Materials		H_2O	366.9		Cl_2	100
		N_aCl	170.5		H_2	2.9
		HCl	0.3		N_aOH	112.6
		H_2SO_4	6.1		H_2SO_4 (dilute)	6.1
		kW h	147			
		H_2O (steam)	620			

(Waste: H_2, N_aOH, H_2SO_4 (dilute))

[a]Inputs listed in (extended) ICGS order, including nontraded resources and wastes.
[b]Per 100 lb of principal product.
Source: Reference A4.

Table 6-A.1(11)   Materials/Energy Input/Output Accounts

Principal product name: **HF**
ICGS Code: **X**

Process or activity name:
Process code: **X**

	Resource/commodity[a] input accounts			Product/waste[a] output accounts		
Column	Extended ICGS code	Name	Input quantity	Extended ICGS code	Name	Output quantity[b]
	1	2	3	4	5	6
Materials		C_aF_2	220		HF	100
		H_2SO_4	279.95		C_aF_2	24.4
					C_aSO_4	345
Energy					H_2SO_4	40.6

} Waste[c]

[a]Inputs listed in (extended) ICGS order, including nontraded resources and wastes.
[b]Per 100 lb of principal product.
[c]Figures estimated from efficiency of given industrial process.
Source: Reference A2.

Table 6-A.1(12)   Materials/Energy Input/Output Accounts

Principal product name: **CHCl₃** ($CHCl_3$)
ICGS code: **X**

Process or activity name: **Methanol process**
Process code: **X**

	Resource/commodity[a] input accounts			Product/waste[a] output accounts		
Column	Extended ICGS code	Name	Input quantity	Extended ICGS code	Name	Output quantity[b]
	1	2	3	4	5	6
Materials		CH_3OH	29.7		$CHCl_3$	100
		HCl	30.5		HCl	57
		Cl_2	118.8		H_2O	15
Energy					CH_3Cl	4
					CH_2Cl_2	3

} Waste[c]

[a]Inputs listed in (extended) ICGS order, including nontraded resources and wastes.
[b]Per 100 lb of principal product.
[c]Figures estimated from efficiency of given industrial process.
Source: Reference A5.

Table 6-A.1(13)   Materials/Energy Input/Output Accounts

Principal product name	CCl₄		Process or activity name	Chlorination of
ICGS code	X		Process code	X methane

Principal product name: CCl_4
ICGS code: X

Process or activity name: Chlorination of
Process code: X | methane

	Resource/commodity[a] input accounts			Product/waste[a] output accounts		
Column	Extended ICGS code	Name	Input quantity	Extended ICGS code	Name	Output quantity[b]
	1	2	3	4	5	6
Energy Materials		CH_4 Cl_2	11.6 184.4		CCl_4 HCl CH_3Cl CH_3Cl_2 $CHCl_3$	100 83.9 2.7 4.3 5.1 } Waste[c]

[a] Inputs listed in (extended) ICGS order, including nontraded resources and wastes.
[b] Per 100 lb of principal product.
[c] Figures estimated from efficiency of given industrial process.
Source: Reference A5.

Table 6-A.1(14)   Materials/Energy Input/Output Accounts

Principal product name: F-22
ICGS code: X

Process or activity name: [ ]
Process code: X

	Resource/commodity[a] input accounts			Product/waste[a] output accounts		
Column	Extended ICGS code	Name	Input quantity	Extended ICGS code	Name	Output quantity[b]
	1	2	3	4	5	6
Energy Materials		$CHCl_3$ HF	138.7 47.2		F-22 HCl F-22 HF	100 83.4 2.3 0.2 } Waste[c]

[a] Inputs listed in (extended) ICGS order, including nontraded resources and wastes.
[b] Per 100 lb of principal product.
[c] Figure estimated from efficiency of given industrial process.
Source: Reference A5.

Table 6-A.2(1) Materials/Energy Origin/Destination Accounts

Material or commodity name: H_2SO_4
ICGS code:

	Origin accounts[a]		Destination accounts[b]					
Originating process code	Quantity produced by each process[d]	Fraction produced by each process	Destination ICGS on ISIC code	Quantity	Fraction[A5]	Conversion process code[c]	Quantity consumed by process[d]	Fraction consumed by process[A2]
1	2	3	4	5	6	7	8	9
P_5	57,000	1.0	2,819	31,350	0.55	P_{10}	2,052	0.036
			2,818	3,990	0.07			
			2,911	4,560	0.08			
			3,323	1,710	0.03	Other destinations not available by process		
			3,356	2,850	0.05			
			2,892	1,425	0.025			
			2,899	285	0.005			
			Misc.	10,830	0.19			

[a]Listed in order of process importance.
[b]Listed as follows: conversions first, in order if ICGS importance (by process, in order of process importance), followed by unconverted uses in order of ISIC importance.
[c]Conversion processes listed where applicable (i.e., where destination is specified by an eight-digit ICGS Code).
[d]Quantity given in thousands of pounds per year.

Table 6-A.2(2) Materials/Energy Origin/Destination Accounts

Material or commodity name: CS_2
ICGS code:

	Origin accounts[a]		Destination accounts[b]					
Originating process code	Quantity produced by each process[d]	Fraction produced by each process	Destination ICGS on ISIC code	Quantity	Fraction[A3]	Conversion process code[c]	Quantity consumed by process[d]	Fraction consumed by process
1	2	3	4	5	6	7	8	9
P_9	479.3	1.0	2281	158.164	0.33	not available by process		
			2818	153.376	0.32	P_7	153.376	0.32
			2921	71.895	0.15			
			Misc.		0.20	not available by process		

[a]Listed in order of process importance.
[b]Listed as follows: conversions first, in order if ICGS importance (by process, in order of process importance), followed by unconverted uses in order of ISIC importance.
[c]Conversion processes listed where applicable (i.e., where destination is specified by an eight-digit ICGS Code).
[d]Quantity given in thousands of pounds per year.

Table 6-A.2(3) Materials/Energy Origin/Destination Accounts

Material or commodity name | Chlorine
ICGS code |

Origin accounts[a] millions lb.			Destination accounts[b]					
Originating process code	Quantity produced by each	Fraction produced by each	Destination ICGS on ISIC code	Quantity	Fraction[A5]	Conversion process code[c]	Quantity consumed by process[d]	Fraction consumed by process
1	2	3	4	5	6	7	8	9
P_2	13,622.	0.66	2,818	12,177.6	0.59	P_6	887.52	0.1
P_1	5,779.2	0.28				P_7	536.64	0.02
P_3	412.8	0.02				P_8	330.24	0.02
P_4	206.4	0.01				P_{12}	6,192.	0.3
Other	619.2	0.03				P_{13}	247.68	0.012
						P_{11}	150.3	0.007
Total	20,640.	1.00	2,611 ⎱ 2,621 ⎰	3,715.2	0.18			
			2,819 ⎱ 4,941 ⎰	2,270.4	0.11		not available by process	
			4,953 ⎱	1,238.4	0.06			
			Misc. ⎰	1,238.4	0.06			

[a] Listed in order of process importance.
[b] Listed as follows: conversions first, in order if ICGS importance (by process, in order of process importance), followed by unconverted uses in order of ISIC importance.
[c] Conversion processes listed where applicable (i.e., where destination is specified by an eight-digit ICGS Code).
[d] Quantity given in thousands of pounds per year.

Table 6-A.2(4) Materials/Energy Origin/Destination Accounts

Material or commodity name | Chloroform
ICGS code |

Origin accounts[a]			Destination accounts[b]					
Originating process code	Quantity produced by each	Fraction produced by each	Destination ICGS on ISIC code	Quantity	Fraction[A5]	Conversion process code[c]	Quantity consumed by process[d]	Fraction consumed[A2] by process[A2]
1	2	3	4	5	6	7	8	9
P_8	126.5	0.5	2818	235.29	0.93	P_{17}	235.29	0.93
P_{11}	126.5	0.5	Misc.	17.71	0.07			

[a] Listed in order of process importance.
[b] Listed as follows: conversions first, in order if ICGS importance (by process, in order of process importance), followed by unconverted uses in order of ISIC importance.
[c] Conversion processes listed where applicable (i.e., where destination is specified by an eight-digit ICGS Code).
[d] Quantity given in thousands of pounds per year.
[e] Dividing the production between the two processes was a simplifying assumption.

202

Table 6-A.2(5) Materials/Energy Origin/Destination Accounts

Material or commodity name	CCl_4
ICGS code	

	Origin accounts[a] millions of lb.		Destination accounts[b]					
Originating process code	Quantity produced by each	Fraction produced by each	Destination ICGS on ISIC code	Quantity	Fraction[A5]	Conversion process code[c]	Quantity consumed by process[d]	Fraction consumed[A2] by process[A2]
1	2	3	4	5	6	7	8	9
P_6	104.7	0.1	2818	1047	1.0	P_{14}	157	0.15
P_7	460.68	0.44				P_{15}	890	0.85
P_8	481.62	0.46						
	1047							
		(e)						

[a]Listed in order of process importance.
[b]Listed as follows: conversions first, in order if ICGS importance (by process, in order of process importance), followed by unconverted uses in order of ISIC importance.
[c]Conversion processes listed where applicable (i.e., where destination is specified by an eight-digit ICGS Code).
[d]Quantity given in thousands of pounds per year.
[e]This is the plant *capacities* production percentage by each process.
[f]0.604 of CCl_4 was for FC-12; 0.396 of CCl_4 was for FC-11.

Table 6-A.2(6) Materials/Energy Origin/Destination Accounts

Material or Comodity Name	HF
ICGS Code	

	Origin accounts[a] millions of lb		Destination accounts[b]					
Originating process code	Quantity produced by each process	Fraction produced by each process	Destination ICGS on ISIC code	Quantity	Fraction[A3]	Conversion process code[c] [A2]	Quantity consumed by process[d]	Fraction consumed by process
1	2	3	4	5	6	7	8	9
P_{10}	733	1.0	2818	293.2	0.406	P_{14}	183.25	0.05
						P_{15}	32.25	0.04
						P_{16}	12.46	0.03
						P_{17}	65.24	0.086
			3334	293.2	0.4			
			2911	29.32	0.04	Not available by process		
			3312	21.99	0.03			
			3339	14.66	0.02			
			Misc.	80.663	0.104			

[a]Listed in order of process importance.
[b]Listed as follows: conversions first, in order if ICGS importance (by process, in order of process importance), followed by unconverted uses in order of ISIC importance.
[c]Conversion process listed where applicable (i.e., where destination is specified by an eight digit ICGS Code).
[d]Quantity given in thousands of pounds per year.

Table 6-A.3　Process Listing

P_1	Chlorine produced by mercury cell
P_2	Chlorine produced by diaphragm cell
P_3	Chlorine produced by Kel-Chlor
P_4	Acetylene produced by Wulff process
P_5	H_2SO_4 by contact process
P_6	Cl_4-C_2Cl_4 (coproduct process) propane chlorination
P_7	Cl_4 carbon disulfide chlorination
P_8	CCl_4-$CHCl_3$ methane chlorination process
P_9	CS_2 by catalytic reaction of sulfur, methane
P_{10}	HF by acidification of fluorspar
P_{11}	$CHCl_3$ my methanol chlorination
P_{12}	EDC by chlorination of ethylene
P_{13}	Perchlorethylene by oxy-chlorination of EDC
P_{14}	F-11/12 vapor phase, fluorination of CCl_4
P_{15}	F-11/12 liquid phase fluorination of CCl_4
P_{16}	F-114, F-113 by fluorination of perchlorethylene
P_{17}	F-22 by fluorination of chloroform

Index